INTRODUCTORY STATISTICAL MECHANICS

David S. Betts and Roy E. Turner
University of Sussex at Brighton

Addison-Wesley Publishing Company

Wokingham, England • Reading, Massachusetts • Menlo Park, California • New York
Don Mills, Ontario • Amsterdam • Bonn • Sydney • Singapore
Tokyo • Madrid • San Juan • Milan • Paris • Mexico City • Seoul • Taipei

© 1992 Addison-Wesley Publishers Ltd.
© 1992 Addison-Wesley Publishing Company Inc.

All rights reserved. No part of this publication may be reproduced, stored in a retrieval system, or transmitted in any form or by any means, electronic, mechanical, photocopying, recording or otherwise, without prior written permission of the publisher.

The programs in this book have been included for their instructional value. They have been tested with care but are not guaranteed for any particular purpose. The publisher does not offer any warranties or representations, nor does it accept any liabilities with respect to the programs.

Many of the designations used by manufacturers and sellers to distinguish their products are claimed as trademarks. Addison-Wesley has made every attempt to supply trademark information about manufacturers and their products mentioned in this book. A list of trademark designations and their owners appears below.

Cover designed by Designers & Partners of Oxford
and printed by The Riverside Printing Co. (Reading) Ltd.
Printed in Great Britain by T.J. Press (Padstow), Cornwall.

First printed 1992.

British Library Cataloguing in Publication Data.
A catalogue record for this book is available from the British Library.

Library of Congress Cataloging-in-Publication Data
Betts, D.S. (David Sheridan),
    Introductory statistical mechanics / by D.S. Betts and R.E. Turner.
       p.   cm.
    Includes bibliographical references and index.
    ISBN 0-201-54421-0 :
    1. Statistical mechanics. I. Turner, R.E. (Roy Edgar),
II. Title.
QC174.5.B47 1992   91-40673
530.1'3 – dc20   CIP

Trademark notice
Vycor is a trademark of Corning Glass Works Incorporated.

Acknowledgement
The publishers wish to thank the following for permission to reproduce material:
Figure 6.5, IOP Publishing Ltd.
Figure 12.3, Oxford University Press.
Figure 12.5, Butterworth-Heinemann Ltd.
Figure 12.10, Copyright © 1978 John Wiley & Sons, Inc.

Preface

In giving lectures on statistical mechanics at Sussex in earlier years we believed that a quiet revolution had taken place in the teaching of physics. This had been signalled by the introduction, principally by Jaynes (1957), of information theory into the formulation of the basic theory of statistical mechanics. From the advantage of looking back now from the 1990s, this belief seems a little over-enthusiastic. However, many people do recognize the simplicity and elegance of this approach and students find it interesting. It is also possible to say now that there has been a revolution at the frontiers of research in areas of relevance to statistical mechanics. In particular, the problem of phase transitions, which had seemed intractable, has been subjected to the power of the renormalization group, with fascinating and revealing consequences.

In the light of these changes, it has been possible to create in this book a clearer presentation. It was intended originally to include a chapter on the third law of thermodynamics, but it proved difficult to present a formally correct argument including verifiable temperature dependences near the absolute zero without going, far beyond the scope of an introductory text.

The book is presented so that general theory is followed by simple applications escalating towards real systems. A significant feature is the chapter on the applications of the renormalization group and Monte Carlo methods to the Ising model. It is not intended that the average undergraduate should be forced to read the later chapters, but they are there for those students who wish to see where the subject has been going in the last twenty years. With these chapters included the book serves as an introductory graduate text. All chapters finish with a summary of the main conclusions.

There are sets of questions at the end of most chapters with solutions where applicable. The use of personal computers by students is now widespread and the student can now easily and quickly obtain accurate data. We have therefore included in the questions examples where the student has to perform numerical calculations.

Finally, at the end of the last chapter we have included a program which shows pictorially a Monte Carlo calculation of a finite two-dimensional Ising model. Even if the student does not want to read the theory of this chapter, he or she is urged to run the program, since we believe it gives considerable enlightenment. The program is interactive in the sense that the student can choose the parameters (non-interacting spins are included as an option). If anything illustrates the change in the application of statistical mechanics, this program does. In the early 1970s, not only was the theory only just beginning, but for the student to have attempted to perform this calculation then, would have meant that it would not be finished by the time this text appeared!

David S. Betts
Roy E. Turner
March 1992

Jaynes E.T. (1957). *Phys. Rev.* **106**, 620 and ibid. (1957) **108**, 171

List of Symbols

Symbol	Name of property or quantity	Page location of first use (also important subsequent use)
a	**Either** a constant in Van der Waals' equation of state	56
	or a length	57, 79
a_s	In Section 12.9, a_s denotes the number of adsorption sites on which s molecules are adsorbed	217
A	Surface area	221
A_j	Amplitude of motion in mode j	80
b	Constant in Van der Waals' equation of state	56
B	**Either** isothermal bulk modulus	39, 52
	or magnetic field	100
B_{int}	Internal magnetic field	247, 260
c	**Either** Heat capacity of sub-system	62
	or Hooke's constant of elasticity	79, 80
	or velocity of wave in a continuous elastic jelly	87
	or speed of light equal to 3×10^8 m sec^{-1}	178
	or a general-purpose constant	232
c_v	Heat capacity at constant volume of sub-system	69
C_V	Heat capacity at constant volume of system under consideration	3
C_B	Heat capacity at constant magnetic field of system under consideration	248

LIST OF SYMBOLS

Symbol	Description	Page
D_e	Interaction potential energy in HD	133
E	Energy, for general purposes	
E_k	Kinetic energy	92
E_n	Energy eigenvalues	11
E_p	Potential energy	92
f	Helmholtz free energy of subsystem: $f = u - Ts$	61
f_{ij}	Function defined in (13.21)	236
F	Helmholtz free energy of system under consideration: $F = U - TS$	43
g	**Either** gyromagnetic ratio	104
	or spin degeneracy $(2S + 1)$	124
g_n	Quantum-mechanical degeneracy of level E_n	42
g_J	Quantum-mechanical degeneracy of level ϵ_J	64
$g(\epsilon)$	Density of energy states	148
$g(p)$	Density of momentum states	176
$g(\omega)d\omega$	Frequency spectrum of oscillating lattice; or of radiation	85, 178
$g_{N\uparrow}$	Degeneracy factor associated with arrangements of atomic moments or spins on a lattice	104
G	Gibbs free energy of system under consideration: **either** $G = U - TS + pV$	50
	or $G = U - TS + MB$ for magnetic systems	103
h	Planck's constant equal to 6.626×10^{-34} Js	29
$\hbar$	Planck's constant divided by 2π, equal to 1.054×10^{-34} Js	40
H	**Either** Hamiltonian operator	11
	or uncertainty	21
	or enthalpy of system under consideration, when $H = U + pV$	196
H_m	Maximum uncertainty	36
I	Moment of inertia	64
j	Quantum number for general purposes	11
J	**Either** a quantum number relating to rotation	65
	or a coupling constant	246
k	Boltzmann's constant, equal to 1.381×10^{-23} J deg^{-1}	40

LIST OF SYMBOLS ix

K	Coupling constant	249
$K(T), K_p(T)$	Equilibrium constants describing degree of dissociation in a diatomic gas	135
L, l, ℓ	**Either** length; or edge of cube	8, 87
	or latent heat	138, 199
m	Mass of atom or particle	79, 80
m,s	See s,m	
m^*	Effective mass (e.g., of electron in metal)	171
m_J	Magnetic quantum number	111
M	**Either** total magnetic moment of system	9, 101
	or number of adsorption sites	217
n	Quantum number;	11
	in 2D or 3D there may be subscripts: n_x, n_y, n_z	67
n_r	Occupation number of single-particle state r	113
n_s	Occupation number of single-particle state s	113
$\{n_s\}$	Set of occupation numbers specifying a microstate	113, 116, 145
N	Total fixed number of atoms (or molecules or particles) in system under consideration	2
N_A, N_B	Numbers of atomic species A and B respectively in binary mixtures	199
N_3, N_4	Numbers of ^{3}He and ^{4}He atoms respectively in ^{3}He/^{4}He mixtures	184
$N_\downarrow, N_\uparrow$	Numbers of down- and up-oriented atomic moments or spins	104
$\mathscr{N}$	Mean number of atoms (or molecules or particles) in system under consideration; $\mathscr{N} = \langle N \rangle$	47
p	**Either** pressure	5, 199
	or electric dipole moment	111, 139
p_A, p_B	Partial vapour pressures of components A and B respectively of a condensed mixture of A and B	199
p^0_A, p^0_B	are the vapour pressures of *pure* condensed phases of A and B respectively	199
p_0, μ, Δ	(occurring together) are parameters	

LIST OF SYMBOLS

	which characterize the excitation spectrum of liquid helium-4	176–7
p_x	x-component of momentum	67
p_y	y-component of momentum	67
p_z	z-component of momentum	67
P	**Either** probability	5
	or electric polarization	111
	or total power emitted per unit area by a black body	182
q	Canonical partition function of sub-system	60
q_e	Electronic canonical partition function of molecule	70
q_r	Rotational canonical partition function of molecule	70
q_t	Translational canonical partition function of molecule	70
q_v	Vibrational canonical partition function of molecule	70
$q(s)$	Canonical partition function for a single adsorption site as sub-system	217
Q	**Either** quantity of heat	15, 37
	or canonical partition function	34, 42
Q_{dist}	Canonical partition function of system composed of distinguishable sub-systems	122
Q_{indist}	Canonical partition function of system composed of indistinguishable sub-systems	122
rms	Root mean square	5
R	Separation of atoms	1
s	**Either** entropy of sub-system	62
	or see s,m	
s,m	In Section 12.9, s denotes the number of molecules adsorbed on a particle site, and m denotes the maximum value of s	217
$\bar{s}$	Mean occupation number for a site s	218
S	**Either** entropy of system under consideration	37
	or spin quantum number	124
T	Temperature in Kelvin	2
T_f	Fermi–Dirac degeneracy temperature	151

T_b	Bose–Einstein degeneracy temperature	157
T_c	Critical temperature for a phase transition	246
T_λ	The temperature of the lambda point transition of liquid helium-4	174
u	**Either** internal energy of sub-system	62
	or phase velocity	81
	or displacement in a continuous elastic jelly	87
u_0	Binding energy of a pair of atoms	198
U	Internal energy of system under consideration; $U = \langle E \rangle$	45, 47
v_A^0	Volume per atom in pure fluid A	209
v_ℓ, v_{t1}, v_{t2}	velocities of longitudinal (ℓ) and transverse (t1 and t2) modes	88
V	Total volume of system under consideration	2
x	**Either** displacement, **or** dimensionless parameter in an integral	84, 221
x_m	Maximum value of x	221
x_n	Displacement of nth atom in chain	79
$x_{nj}(t)$	Displacement as a function of time of nth atom in jth mode	80
X	Atomic fraction of ^{3}He in ^{3}He/^{4}He mixtures	184
X_3	Atomic fraction of ^{3}He in ^{3}He/^{4}He mixtures	205
X_A, X_B	Atomic fractions of components A and B in a binary mixture	199
z	Number of nearest neighbours in a lattice	198, 204
Z	A thermodynamic potential, $Z = \mu \mathcal{N} + TS - U$	50, 146
Z_N	Classical configuration integral	234
α	**Either** dissociated fraction (in gas of diatomic molecules)	134, 141
	or dimensionless parameter in relationship $\epsilon_s = V^{-\alpha}$	14, 148
β	**Either** magnetic moment of the electron	168
	or undetermined (Lagrangian) multiplier;	28
	also in systems, $\beta = 1/kT$	43

xii LIST OF SYMBOLS

Symbol	Description	Page
γ	**Either** undetermined (Lagrangian) multiplier	48
	or magnetic moment of atom;	104, 211
	also in systems, $\gamma = -\mu/kT$	50
$\delta_{jj'}$	Delta function equal to unity if $j = j'$, otherwise zero.	92
Δ, p_0, μ	(occurring together) are parameters which characterize the excitation spectrum of liquid helium-4	176–7
ϵ	Energy parameter in Lennard–Jones atomic interaction;	2, 242
	or more generally	57, 114
ϵ_F, ϵ_f	Fermi energy equal to kT_f	168, 211
ϵ_e	Electronic energy of molecule	69
ϵ_r	Rotational energy of molecule	69
ϵ_t	Translational energy of molecule	69
ϵ_v	Vibrational energy of molecule	69
$\epsilon_\downarrow, \epsilon_\uparrow$	Energies of down- and up-oriented atomic moments or spins	104
ϵ_0	**Either** single-particle ground state energy	157, 158
	or permittivity of free space equal to 8.854×10^{-12} farad m^{-1}	173
$\epsilon_3(X_3)$	Binding energy of a ^{3}He atom in a predominantly liquid-^{4}He environment	206
ζ	Parameter used in Chapter 14	250
$\zeta(n)$	Dimensionless Riemann zeta function	158
θ_v	Characteristic temperature for a simple harmonic oscillator	64
θ_r	Characteristic temperature for a rotator	65
θ_t	Characteristic temperature for translational motion	67
θ_D	Characteristic temperature in Debye model	89
θ_E	Characteristic temperature in Einstein model	76
θ_M	Characteristic temperature for magnetic solids	105
θ_0	A characteristic temperature indicating the onset of degeneracy in a gas of fermions or bosons	125

LIST OF SYMBOLS   xiii

λ	**Either** undetermined (Lagrangian) multiplier	28
	or wavelength	80
	or (Section 12.9 only) $\exp(\mu/kT)$	218
λ_{min}	Minimum wavelength of wave in lattice	89
Λ_i	A quantum length parameter	138
μ	**Either** atomic magnetic moment	9
	or chemical potential	17, 192
μ_A, μ_B	Chemical potentials of atomic species A and B respectively in binary mixtures	199
μ^0_A, μ^0_B	are the chemical potentials of atomic species A and B respectively in their *pure* phases	200
μ_s, μ_l, μ_g	are chemical potentials of solid, liquid, and gas phases respectively	198
μ, Δ, p_0	(occurring together) are parameters which characterize the excitation spectrum of liquid helium-4	176–7
μ_0	**Either** magnetic permeability of free space, equal to $4\pi \times 10^{-7}$ henry m^{-1}	111
	or the chemical potential of a perfect gas of fermions as $T \to 0$	151
μ_B	Bohr magneton, equal to 9.274×10^{-24} J T^{-1}	104
$\mu_3(3)$	Chemical potential of a ^{3}He atom in liquid ^{3}He	206
$\mu_3(4)$	Chemical potential of a ^{3}He atom in a predominantly liquid-^{4}He environment	205
ν	Frequency (or cycles per second) in Hertz; $\nu = \omega/2\pi$	29
ξ	A factor, corresponding to an adsorption site, of the grand canonical partition function Ξ	218
ξ_j	Normal mode displacement co-ordinates	82
Ξ	Grand canonical partition function	48
Ξ_f	Grand canonical partition function for a gas of fermions	145
Ξ_b	Grand canonical partition function for a gas of bosons	146
Π	Osmotic pressure	184–5, 208

LIST OF SYMBOLS

Symbol	Description	Page
ρ	Number density, $\rho = N/V$	237, 243
σ	Distance parameter in atomic interaction	2, 242
φ	Interaction energy	1
$\varphi(r_{ij})$	Interaction energy as a function of separation r_{ij} of two atoms	236
$\varphi(\mu, T)$	A function used in discussion in ...	51
φ_j	Phase angle of motion in mode j	80
Φ	**Either** work function	214
	or potential energy	234
Φ_1, Φ_2	General-purpose functions	226
Φ_n	Probability of system having energy E_n	11
χ	Magnetic susceptibility	111
Ψ, ψ	Eigenfunction	11, 119
ω	Angular frequency (or speed) in rad/sec: $\omega = 2\pi\nu$	40
ω_j	Angular frequency of mode j	80
ω_m	Maximum angular frequency in models of solids	80, 85, 88
Ω	Total number of microstates; microcanonical partition function	22
$\langle X \rangle$	Mean value of X	8
$\langle \Delta X \rangle_{\text{rms}}$	Root mean square deviation. $\langle \Delta X \rangle_{\text{rms}}^2$ is equal to $\langle X^2 \rangle - \langle X \rangle^2$	6

Contents

Preface		v
List of Symbols		vii
Chapter 1	**Introduction**	1
	1.1 The problem of statistical mechanics	1
	Questions	7
Chapter 2	**The Conservation of Energy**	10
	2.1 Introduction	10
	2.2 Reversible changes and the conservation of energy	12
	2.3 Variable number of particles	16
Chapter 3	**Statistical Inference**	19
	3.1 Introduction	19
	3.2 Uncertainty	20
	3.3 A scale of uncertainty	21
	3.4 Equal probabilities	22
	3.5 Rational requirements for the function H	22
	3.6 The function H which satisfies conditions A, B and C	23
	3.7 Unequal probabilities	24
	3.8 Strategy of least bias	25
	3.9 Case 1: toss of a coin	26
	3.10 Case 2: N atoms each with three energy levels	26
	3.11 Case 3: a loaded die	27
	3.12 Case 4: an assembly of $3N$ simple harmonic oscillators	29
	Questions	32

xvi   CONTENTS

Chapter 4	**The Law of Maximum Uncertainty**	**35**
4.1	Introduction	35
4.2	Thermodynamically isolated system (microcanonical distribution)	35
4.3	System in a heat bath (canonical distribution)	40
4.4	System in a heat-and-particle bath (grand canonical distribution)	47
	Questions	54
Chapter 5	**Partition Functions of Simple Sub-systems**	**59**
5.1	Introduction	59
5.2	Particle with two possible states with energies ϵ_1 and ϵ_2	61
5.3	One-dimensional simple harmonic oscillator (SHO)	63
5.4	Rotating heteronuclear diatomic molecule	64
5.5	Particle-in-a-box	67
5.6	One heteronuclear diatomic molecule in a box	69
5.7	Two weakly interacting distinguishable molecules in a box	70
5.8	Conclusion	71
	Questions	72
Chapter 6	**Models of Simple Solids**	**74**
6.1	Introduction	74
6.2	The Einstein model	75
6.3	A one-dimensional oscillating lattice model	79
6.4	A three-dimensional oscillating lattice model	85
6.5	The Debye model	87
6.6	An alternative approach	90
	Questions	92
	Reference	99
Chapter 7	**Models of Magnetic Crystals**	**100**
7.1	Introduction	100
7.2	Magnetic systems in the canonical distribution	101
7.3	Magnetic systems in the grand canonical distribution	102
7.4	Paramagnetic crystal in the canonical distribution	103
7.5	Paramagnetic crystal in the grand canonical distribution	108
	Questions	110
Chapter 8	**Systems Composed of Non-interacting Sub-systems**	**113**
8.1	Many-particle states	113
8.2	Distinguishable particles	116

			CONTENTS	xvii

	8.3	Indistinguishable particles	118
		Question	120

Chapter 9 Aspects of the Classical Gas — 122

	9.1	The high-temperature low-density criterion	122
	9.2	The thermodynamic properties of a monatomic gas	128
	9.3	Thermodynamic properties of polyatomic gases	129
	9.4	Non-reacting gas mixtures	132
	9.5	Reacting gas mixtures and dissociation	132
	9.6	Maxwell-Boltzmann energy distribution	136
		Questions	137
		References	143

Chapter 10 Gases in the Grand Canonical Distribution — 144

	10.1	General formulation	144
	10.2	The classical gas and the Maxwell-Boltzmann distribution	148
	10.3	The Fermi gas	150
	10.4	Fluctuations in a fermion system	155
	10.5	The Bose gas	157
	10.6	Fluctuations in Bose systems	163
		Questions	165
		References	168

Chapter 11 Real Fermi and Bose systems — 169

	11.1	Introduction	169
	11.2	The 'gas' of conduction electrons in metals	170
	11.3	Liquid helium-4	174
	11.4	Electromagnetic radiation	178
	11.5	Helium-3, the Fermi fluid	183
		Questions	187
		References	191

Chapter 12 Equilibrium in Two-phase Assemblies — 192

	12.1	The chemical potential	192
	12.2	Vapour pressure of an Einstein solid	197
	12.3	The ideal mixture	199
	12.4	Phase separation of mixtures	203
	12.5	Osmotic pressure	208
	12.6	Electron paramagnetism	211
	12.7	Contact potential	214
	12.8	Surfaces	216
	12.9	Adsorption isotherms	217
	12.10	Two-dimensional solid layers	219
	12.11	Two-dimensional gaseous layers	223
		Questions	226
		References	229

Chapter 13	The Interacting Classical Gas		230
	13.1	Classical statistical mechanics	230
	13.2	The interacting gas	234
	13.3	Evaluation of the second virial coefficient	238
		Questions	241
		References	244
Chapter 14	Modern Approaches to Phase Transitions		245
	14.1	Introduction	245
	14.2	The Ising model	246
	14.3	The mean field approximation	247
	14.4	The renormalization group and the one-dimensional Ising model	249
	14.5	The renormalization group and the two-dimensional Ising model	252
	14.6	Monte Carlo methods	255
		Questions	263
		References	268

Appendices

Appendix 1	The method of Lagrange or undetermined multipliers	269
	Questions	271
Appendix 2	Number fluctuations in the grand canonical distribution	272
Appendix 3	Proof that, for independent distinguishable sub-systems, $Q = \Pi q$	275
Appendix 4	Tables of the chemical potential and energies for Fermi and Bose gases	276
	Question	278
Appendix 5	Integrals for the Fermi gas	278

Numerical answers to questions   282

Index   285

CHAPTER 1
Introduction

1.1 The problem of statistical mechanics

Questions

1.1 The problem of statistical mechanics

The type of problem to which statistical mechanics can be applied is best conveyed by considering an example. The reader may imagine a block of solid argon upon which, inside a suitable low-temperature refrigerator, a variety of experiments can be performed. We could start by performing an X-ray scattering experiment, which would tell us that the crystal structure is face-centred cubic with a nearest neighbour distance of 0.37 nm. From this (and the known volume of the crystal) can be calculated the number of argon atoms in the sample. We could also measure the cohesive energy of the crystal; this is approximately the energy needed to turn the solid into a vapour. The fact that we have to supply energy implies that the force between two argon atoms must have an attractive part. Since the atoms have a finite separation, the forces between them must also have a repulsive part. Next, we could measure the elastic properties of the crystal and find the energy needed to squeeze and stretch it. This will tell us something about the magnitude of the forces in the region of the nearest neighbour distance. All this data is reasonably consistent with the following expression for the interaction energy (φ) for a pair of atoms at a separation R,

$$\varphi(R) = 4\epsilon \left[\left(\frac{\sigma}{R}\right)^{12} - \left(\frac{\sigma}{R}\right)^{6} \right] \tag{1.1}$$

where σ ($= 0.34$ nm) and ϵ ($= 1.67 \times 10^{-21}$ joules) are constants.

At this point it is formally possible to write down a Schrödinger equation for the crystal and solve it to find the eigenstates and eigenvalues. (In practice this may be very difficult and we may have to resort to approximate solutions. This does not, however, affect the logic of the argument.) Finally, let us perform one more experiment, the measurement of heat capacity as a function of temperature. The results of this experiment will show that the heat capacity is closely proportional to T^3 at temperatures close to 1 K and very nearly constant at a temperature around 84 K (the normal melting point). Why is this, and does it follow automatically from the solution of the Schrödinger equation?

The problem may be stated in more general terms: how is it possible, given a microscopic model Hamiltonian, to predict the macroscopic behaviour of the system? Several difficulties can be listed which will clearly arise between the statement of the problem and its solution. Firstly, how important are the boundary conditions? In the above example we imagined the system (crystal) in a heat bath (refrigerator) but we would expect the same macroscopic result if the system was allowed into equilibrium with the heat bath and then isolated from it. These two situations require different microscopic descriptions. Secondly, which eigenstate of the system, out of the infinite possible states, is the system actually in? Indeed, is it correct to speak of the system being in an eigenstate at all? To determine the exact state of the system it would be necessary to measure all the quantum numbers (which would ensure that after the measurement, the system would be in an eigenstate). The total number of measurements in our example would have to be at least $3N$, where N is the number of atoms, N itself being of the order of 10^{24}. In practice we can usually make only a few measurements to determine the state of the system, for instance, the number of atoms N and volume V (which between them determine the *possible* eigenstates of the system) and the temperature T. This latter measurement does not have an obvious microscopic interpretation.

Let us review the situation. We know the possible states of the system, derived from the solutions of the Schrödinger equation. In determining the actual state of the system we need of the order of 10^{24} measurements, whereas in practice we have only one, the temperature. This last sentence makes it clear that what we are faced with is no longer a problem in physics but a problem of statistical inference. We know the possible states (or more precisely, the accessible states) of the system, but we have only one piece of information to determine which of the accessible states it is in. What can rationally be inferred from this about the behaviour of the system?

It is perhaps worth noting at this point that we have chosen to talk about the temperature as our one piece of information: this is because, experimentally, this is what is usually done. There is no reason, however, why we should not measure the total energy of the system instead of the temperature. This can be done, for example, by measuring the heat capacity at constant volume C_V as a function of temperature. Since $C_V = (\partial U/\partial T)_V$ it follows that if C_V is plotted as a function of temperature, the area under the curve from $T = 0$ to the final temperature of the system will give the internal energy of the system (apart from an additive constant which defines the energy of our system at absolute zero). This measurement, together with the number and volume, will again uniquely determine the macroscopic state (see Chapter 4). Since the total energy is easier than temperature to interpret from a microscopic point of view, we shall for the moment take this as our given piece of information.

It is clear that in making predictions in statistical mechanics we are perforce dealing not with predictive certainties, but with probabilistic inferences made against a background of considerable ignorance. The concept of probability is here taken to refer to a quantifiable and rational scale of confidence or belief in the outcome of an experiment, and the assignment of a numerical probability has to be made *in spite of uncertainty* about many relevant factors, namely all the quantum numbers. If there were no such missing information, we should be dealing with a predictive certainty. Incidentally, in some everyday examples of assignments of probabilities, the numerical assignments are testable in a limited sense, because trials can be repeated many times, in which case probabilities are closely related to relative frequencies. This would apply to, for instance, rolling dice. However, some situations are not readily repeatable without altered parameters, for instance, guessing the likelihood of passing an examination or the wisdom of making a U-turn on a main road. It is to cover all these situations that we have referred to a scale of confidence. The requirements of rationality might be satisfied by repetition or otherwise – more of this later.

In statistical mechanics a suitable approach might be: from the given information, we would like to predict what the probability is of finding the system in a particular microscopic state, i.e., an eigenstate. To verify whether our predictions are correct we would have to take a large number of identical systems in the same macroscopic state and measure the exact microscopic state of each of them. The number of times a particular microscopic state occurred divided by the total number of identical systems should be approximately equal to the predicted probability. In the limit of the number of identical systems tending to infinity the equality should no longer be approximate but exact. From this argument it is clear that if we did predict the probability of a particular microscopic state it would be experimentally completely unverifiable.

It is therefore worthwhile to examine more closely what we do wish to predict, our criteria being that it should be possible both in principle and in practice to have experimental verification. Now in practice we are interested not in 10^{24} measurements on a system, but something less than 10, for

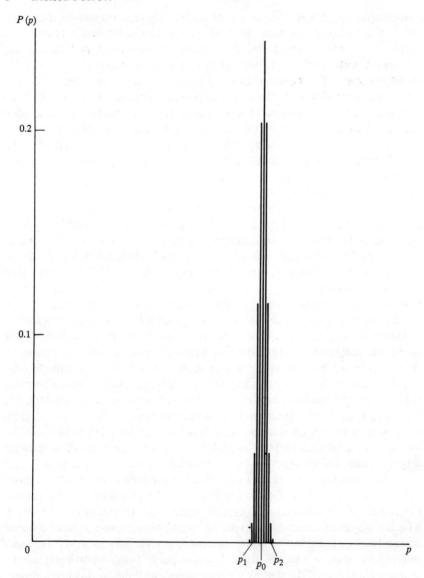

Figure 1.1 A schematic plot of the probability $P(p)$ of a system having a particular pressure p. The most likely pressure is p_0 and it is most unlikely that a pressure less than p_1 or greater than p_2 be observed. Observation of macroscopic real systems suggests that $(p_2 - p_1) \ll p_0$, i.e., the distribution of probabilities is so sharp that p_0 may be taken as the observed value and identified with the mean value $\sum_p pP(p)$.

example, heat capacities, equation of state, etc. Furthermore, we are only likely to repeat these measurements on a very few identical systems and not anywhere near a sufficiently large number for it to be regarded as approximately infinite. Finally, and very importantly, however big or small the number of identical systems is, we expect to get, within the experimental error, the same results for our macroscopic measurements for each system. In other words, our statistical inference should lead to predicting macroscopic measurements with virtual certainty, since we know that the behaviour of real systems is highly predictable.

From this analysis the following points become clear. Firstly, what we need to predict are not the probabilities of individual states, but the probabilities of macroscopic quantities occurring, for example the probability of a particular pressure being observed. Secondly, the form of these probabilities must be such that they are strongly peaked about a particular value and outside this peak the probability is practically zero. Furthermore, the values of the macroscopic parameter for which the probability is significant should be indistinguishable from the point of view of macroscopic measurements. For it is only in this way that we can ensure that our results will always be reproducible on a macroscopic scale. A schematic plot of the probability $P(p)$ of a system having a particular pressure p is shown in Figure 1.1.

Clearly, since all the values of the parameters where the probability is significant are identical macroscopically, we can choose any one of them to represent the observed value. In particular it is often convenient to choose the mean value. This is purely for convenience since we could equally well choose some other value, for example the value where the probability is a maximum. On the macroscopic scale this value would be indistinguishable from the mean value. (If the probability distribution is symmetric about its maximum value, then the mean value and the value where the probability is a maximum would be identical, even on a macroscopic level. This, however, does not affect our general argument that whatever the form of the distribution function, the two values are macroscopically indistinguishable.) As we have said, this argument depends on the width of the probability function being vanishingly small on the macroscopic scale. If this were not so, then the probabilities for, say, two macroscopically distinguishable values of the parameters would be significant. Thus in any experiment, we would have a finite chance of getting one of two values for the result. A convenient measure of the width of the peak is the root mean square deviation (r.m.s. deviation). Thus we have to show that the r.m.s. deviation divided by the mean value (identical to the observed macroscopic value) is vanishingly small on a macroscopic scale, say of order $1/N$ where N ($\sim 10^{24}$) is the number of particles in the physical system. We can illustrate these ideas by returning to our example of the probability function for pressure given in Figure 1.1.

Identifying the observed value of the pressure with the mean value, we have

$$p_{\text{obs}} = \sum_p P(p)p = \langle p \rangle = p_0 \qquad (1.2)$$

The r.m.s. deviation is given by

$$\begin{aligned}\langle\Delta p\rangle^2_{\text{rms}} &= \sum_p P(p)(p-p_0)^2 \\ &= \sum_p P(p)(p^2 - 2pp_0 + p_0^2) \\ &= \langle p^2\rangle - 2\langle p\rangle p_0 + p_0^2 \\ &= \langle p^2\rangle - p_0^2 \end{aligned} \qquad (1.3)$$

since the normalization conditions requires that

$$\sum_p P(p) = 1 \qquad (1.4)$$

In the present case, reference to Figure 1.1 shows that the fractional r.m.s. deviation is given approximately by

$$\frac{\langle\Delta p\rangle_{\text{rms}}}{p_0} \simeq \frac{p_2 - p_1}{p_0} \qquad (1.5)$$

On the macroscopic scale this must be vanishingly small, otherwise p_0, p_1 and p_2 (and any values between p_1 and p_2) would be possible outcomes of a macroscopic experiment.

Thus we have reduced our original formidable problem of inferring the exact probabilities of microscopic states, to the much simpler one of inferring the mean and root mean square deviation of a small number of macroscopic observables. Now it will turn out later on that, in the course of deducing these statistical averages, we will be able to write down the probability for the system being in a particular microscopic state but, as far as statistical mechanics is concerned, we will not know whether it is 'correct' or not. This is because the only test we have of 'correctness' is whether that distribution predicts mean and mean square deviations of macroscopic variables which are in agreement with experiment. If we had used a different approach and derived different probability functions but the same values for the macroscopic variables, then the new probability function would have been equally as 'correct' as the previous one. We shall see in later chapters that by assuming different boundary conditions we will find different probability functions but, with the exception of one interesting case, the same values of the macroscopic variables. (See the remarks above about the effect of boundary conditions.) In the next chapter, we will see that we can make some progress by simply assuming that the mean values exist but not actually calculating them.

Finally, the remarks we have made so far, with the exception of the paragraph above, would apply equally well to the so-called transport or non-equilibrium processes. However, we shall here restrict ourselves to systems where *all* the macroscopic parameters are independent of time and independent of the position in the sample, for example, pressure and magnetic field are uniform throughout the sample. This latter condition excludes the possibility of steady-state flow processes which would not be excluded by the time-independence criterion alone. For example, the flow of a liquid through a tube with a constant pressure difference would eventually arrive at a time-independent steady state, but there would be a pressure gradient down the tube which would exclude it from our considerations. The systems we will be considering are said to be in thermodynamic equilibrium.

CHAPTER 1: SUMMARY OF MAIN POINTS

1. Introduction to the central problem of statistical mechanics: how to deduce macroscopic properties of a material from a detailed knowledge of the behaviour of its microscopic (atomic) constituents.
2. Very large numbers of microscopic constituents reduce the root mean square deviations.
3. For example, the probability of measuring a gas pressure which is more than a little removed from the average value is extremely small.

QUESTIONS

1.1 Suppose the probability density for the speed s of a car on a highway is given by

$$\rho(s) = As \exp\left(-\frac{s}{s_0}\right) \quad \text{where } 0 \leqslant s \leqslant \infty$$

and A and s_0 are positive constants. More explicitly, $\rho(s)\,ds$ gives the probability that a car has a speed between s and $s + ds$.

(a) Determine A in terms of s_0.
(b) What is the mean value of the speed?
(c) What is the 'most probable' speed, that is, the speed for which the probability density has its maximum? Compare this with the expectation value.
(d) What is the probability that a car has a speed more than three times as large as the mean value?

1.2 The displacement x of a classical simple harmonic oscillator as a function of the time t is given by

$$x = A\cos(\omega t + \varphi)$$

where ω is the angular frequency of the oscillator, A the amplitude of oscillation, and φ is an arbitrary constant which can have any value in the range $0 \leqslant \varphi \leqslant 2\pi$. Find the probability $\rho(x)dx$ that the displacement of the oscillator, at a random instant, is found to lie in the range between x and $x + dx$.

1.3 A chain consists of N links each of length a. Each link either increases the length of the chain by a or is folded back on its preceding neighbour and thus does not increase the length. If we denote the two possible configurations by a_α for the αth link in the chain, the total length is

$$L = \sum_\alpha a_\alpha$$

where the possible values of a_α are $a_\alpha = +a$ or $a_\alpha = -a$. If we denote by p (the same for all links) the probability that for the αth link $a_\alpha = +a$ and the probability of $a_\alpha = -a$ by $q = 1 - p$, show that the average length of the chain is

$$\langle L \rangle = N\langle a_\alpha \rangle$$

where $\langle a_\alpha \rangle = pa - qa = (2p-1)a$.

Similarly the square of the length of the chain is

$$L^2 = \sum_\alpha \sum_\beta a_\alpha a_\beta = \sum_\alpha a_\alpha^2 + \sum_{\alpha \neq \beta} a_\alpha a_\beta$$

Hence by using the fact that probabilities for the configurations of two links are independent, i.e., the probability for a pair of links α, β having a configuration $a_\alpha a_\beta$ is simply the product of probability for the αth link having a configuration of a_α and the probability for the βth link having a configuration a_β, show that the average value of the square of the length is

$$\langle L^2 \rangle = N\langle a_\alpha^2 \rangle + N(N-1)\langle a_\alpha \rangle^2 \quad \text{where } \langle a_\alpha^2 \rangle = a^2$$

For the case of large N show that the probability for the chain having a length L is sharply peaked about $\langle L \rangle$.

This suggests that (for this case at least) large numbers sharpen distributions. The analysis can readily (with minor changes in notation) be used to describe diffusion in a lattice (atoms move to left or right), rubber elasticity (links in polymer chain in forward or reverse configuration) or a paramagnetic solid (each magnetic ion may be in spin-up or spin-down configuration).

1.4 Consider a nucleus having spin 1 (i.e., spin angular momentum $\hbar$). Its component μ of magnetic moment along a given direction can then have three possible values, namely $+\mu_0$, 0, or $-\mu_0$. Suppose that the nucleus is not spherically symmetrical, but is ellipsoidal in shape. As a result it tends to be preferentially oriented so that its major axis is parallel to a given direction in the crystalline solid in which the nucleus is located. There is thus a probability p that $\mu = +\mu_0$, and a probability p that $\mu = -\mu_0$; the probability that $\mu = 0$ is then equal to $1 - 2p$.

(a) Calculate $\langle \mu \rangle$ and $\langle \mu^2 \rangle$.

(b) Calculate $\langle \Delta \mu \rangle^2_{rms}$.

(c) Suppose that the solid under consideration contains N such nuclei which interact with each other to a negligible extent. Let M denote the total component of magnetic moment along the specified direction of all these nuclei. Calculate M and its r.m.s. deviation M_{rms} in terms of N, p, and μ_0.

1.5 Consider an ideal system of N identical spins $\frac{1}{2}$. The number n of magnetic moments which point in the up direction can then be written in the form $n = \mu_1 + \mu_2 + \ldots + \mu_N$, where $\mu_i = 1$ if the ith magnetic moment points up and $\mu_i = 0$ if it points down. Use the given expression for n and the fact that the spins are statistically independent to establish the following results.

(a) Show that $\langle n \rangle = N \langle \mu \rangle$.

(b) Show that $\langle \Delta n \rangle^2_{rms} = N \langle \Delta \mu \rangle^2_{rms}$

(c) Suppose that a magnetic moment has a probability p of pointing up and a probability $q = 1 - p$ of pointing down. What are $\langle \mu \rangle$ and $\langle \Delta \mu \rangle^2_{rms}$?

(d) Calculate $\langle n \rangle$ and $\langle \Delta n \rangle^2_{rms}$.

1.6 The probability of finding n photons of a given frequency in an enclosure which is in thermal equilibrium with its walls is

$$P_n = (1 - \alpha) \alpha^n$$

where $0 \leqslant \alpha \leqslant 1$ and α is a function of the frequency, the volume of the enclosure and the temperature of the walls. What is the mean number $\langle n \rangle$ of photons of this frequency? Show that the fractional deviation is given by

$$\frac{\langle \Delta n \rangle_{rms}}{\langle n \rangle} = \left(1 + \frac{1}{\langle n \rangle} \right)^{\frac{1}{2}}$$

CHAPTER 2
The Conservation of Energy

2.1 Introduction
2.2 Reversible changes and the conservation of energy

2.3 Variable number of particles

2.1 Introduction

Our discussion in the last chapter attempted to show that the problem of statistical mechanics is to predict the macroscopic properties of the system from its microscopic properties. By 'system' we mean the actual physical system placed, for example, on the laboratory bench or in a cryostat. Henceforth in our discussion the word system, will always mean this. Its macroscopic state will be defined by a small number of variables, for example, volume, number of particles, pressure, etc. Thus, if we have, say, two blocks of the same material at the same pressure and volume (and since they both have the same density, the same number of particles), then they will both be in the same state. The possible microscopic states of the system will be the eigenstates of the time-independent Schrödinger equation. Strictly speaking this is not the whole story, since in general the system will be in a linear combination of eigenstates but we are assuming that we can make a hypothetical experimental determination of its eigenstate. This measurement will according to quantum mechanics ensure that the system is driven into an eigenstate. We can then ask the question: what is the probability that we will get a particular result? Again, strictly speaking, we cannot know that the measurement of the eigenstate will not alter the macroscopic properties of the system. This would, however, take us into deeper waters than is warranted in an introductory text. We shall therefore assume that the macroscopic properties are not affected by any such measurements. We should, however, distinguish between when the system is in a heat bath, when it has semi-permeable walls and when it is totally isolated, for these boundary conditions, in principle, affect the eigenstates and eigenvalues. However, whatever the boundary conditions, we can, according to our previous discussion, assume that there exists a probability distribution for the total energy and equate the mean value to the observed value of the energy.

Let us suppose that the microstates (i.e., the solutions of Schrödinger's equation) can be labelled by an *energy* quantum number n and the remaining set of quantum numbers by the symbol $\{j\}$. Then the Schrödinger equation may be written

$$H\Psi_{n\{j\}} = E_n \Psi_{n\{j\}} \tag{2.1}$$

where H is the Hamiltonian of the system, $\Psi_{n\{j\}}$ are the eigenfunctions and E_n the energy eigenvalues. If $P_{n\{j\}}$ is the probability that the system is in the state $(n, \{j\})$, then the mean value of the energy is given by

$$U = \sum_n \sum_{\{j\}} P_{n\{j\}} E_n = \sum_{\{n\}} \Phi_n E_n \tag{2.2}$$

where

$$\Phi_n = \sum_{\{j\}} P_{n\{j\}} \tag{2.3}$$

is the probability of the system having an energy E_n. This mean energy U is to be equated to the observed value of the energy. The first question we can ask is: what macroscopic variables, possibly amongst others, would we expect U, as defined by equation (2.2), to depend on? Clearly this dependence will come either through the Φ_n or through E_n. At this stage we know nothing about the Φ_n beyond the assumption that it is sharply peaked about its mean value. We can, however, infer something about the dependence of the energy eigenvalues E_n from the Schrödinger equation (2.1). Typically H has the form

$$H = \sum_{\alpha=1}^{N} \frac{p_\alpha^2}{2m} + V(r_1, r_2, r_3, \ldots r_\alpha \ldots r_N) \tag{2.4}$$

where p_α is the momentum operator for the αth particle and r_α is its position vector. The first term is the sum of the kinetic energies of the N particles and the second term the potential energy of their mutual interaction. Thus

12   THE CONSERVATION OF ENERGY

we would expect the energy eigenvalues E_n to depend on N. Furthermore, we have the normalization condition on the eigenfunctions

$$\int_V dr_1 \ldots \int_V dr_2 \ldots \int_V dr_\alpha \ldots \int_V dr_N \Psi^*_{n\{j\}} \Psi_{n\{j\}} = 1 \qquad (2.5)$$

where each of the N integrations is taken over the volume of the system. Since an alternative expression for the eigenvalues is

$$E_n = \int_V dr_1 \ldots \int_V dr_2 \ldots \int_V dr_\alpha \ldots \int_V dr_N \Psi^*_{n\{j\}} H \Psi_{n\{j\}}, \qquad (2.6)$$

we might expect the E_n to depend on the volume. We should emphasize that this argument is not conclusive but merely intended to show that in general it is reasonable to expect the E_n to depend on N and V. But it does not exclude the possibility of the E_n depending on, say, temperature (although as we have already remarked, temperature does not have an obvious microscopic intepretation).

Next we make use of a physical law that is reliable both at the microscopic and macroscopic level. This is the law of the conservation of energy. Its interpretation at the microscopic level is well known as indeed is its statement at the macroscopic level in the form of the first law of thermodynamics. Thus, if we consider changes in a system, we can examine their effects (on the total energy) at both levels and hopefully get some further insight into the connection between the two physical domains.

2.2   Reversible changes and the conservation of energy

In thermodynamics a change in a system is said to be reversible if it is carried out infinitely slowly. To carry out such a change we alter one of the parameters of the system, for example, the pressure. As a consequence of changing this parameter, other parameters of the system will change infinitely slowly, for example, altering the pressure may produce changes in volumes or temperature or both.

What, then, is the microscopic interpretation of such a change? Clearly what we are doing is applying a perturbation to the system infinitely slowly. If we make the change very small, then we can apply time-dependent perturbation theory to the problem.

Now let us assume that our physical system is isolated and subjected to a small infinitely slow volume change δV (since the system is isolated the change is said to be an adiabatic reversible one), for example, a small compression of a gas by a piston. The energy eigenvalues will change from E_n to E_n' where

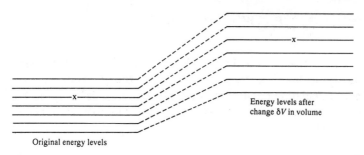

Figure 2.1 Schematic representation o the effect of applying an *adiabatic* perturbation (in this case a volume change δV). The energy levels shift but the system (represented by a cross) remains in the corresponding state. If the perturbation were not adiabatic, then the shifts in the energy levels would be the same but there would be no guarantee that the system would remain in the corresponding state.

$$E_n' = E_n + \left(\frac{\partial E_n}{\partial V}\right)_N \delta V \tag{2.7}$$

Furthermore, the time-dependent perturbation theory tells us that, if the system was initially in the state $(n, \{j\})$, it will go to the corresponding eigenstate when the perturbation has been applied. (The proof of this statement is a straightforward application of time-dependent perturbation theory; see, for example, L. I. Schiff, *Quantum Mechanics* (McGraw-Hill, 1955).) This situation is illustrated in Figure 2.1. We can express this in another way by noting that, in general, time-dependent perturbation theory gives us two results, namely (a) the change in the energy levels and (b) the transition probability, i.e., the probability that, given the system in an initial state, it makes a transition to a different final state. For the case where the perturbation is applied infinitely slowly this latter transition probability is zero for all final states except where the final state is the same as the initial state, where it is equal to one. This then describes the microscopic picture; the question is now what happens macroscopically?

We do not know what state our actual physical system is in initially, only the probability $P_{n\{j\}}$ that the system is in a particular state. There exists presumably also a probability $P'_{m\{k\}}$ that the system is finally in a state $(m, \{k\})$. We can calculate $P'_{m\{k\}}$ as follows. The probability that the system is in a final perturbed state $(m, \{k\})$ will be equal to the probability that the system was initially in the state $(n, \{j\})$ times the probability that, given the system in state $(n, \{j\})$, it makes a transition to the state $(m, \{k\})$ and then is summed over all initial states $(n, \{j\})$. In mathematical language this argument may be written

14   THE CONSERVATION OF ENERGY

$$P'_{m\{k\}} = \sum_n \sum_{\{j\}} P_{n\{j\}} W(n, \{j\} | m, \{k\}) \tag{2.8}$$

where $W(n, \{j\} | m, \{k\})$ is the transition probability from the given initial state $(n, \{j\})$ to the final state $(m, \{k\})$. Now the results of our perturbation theory calculation tell us that $W(n, \{j\} | m, \{k\})$ is zero unless $(n, \{j\})$ and $(m, \{k\})$ are the same state. Applying this result to equation (2.8) gives us immediately

$$P'_{m\{k\}} = P_{m\{k\}} \tag{2.9}$$

since $W(m, \{k\} | m, \{k\}) = 1$. We are now in a position to calculate the new perturbed mean energy, using equations (2.7) and (2.9),

$$U' = \sum_n \sum_{\{j\}} P'_{n\{j\}} E'_n$$

$$= \sum_n \sum_{\{j\}} P_{n\{j\}} \left[E_n + \left(\frac{\partial E_n}{\partial V}\right)_N \delta V \right]$$

Hence the change in the mean, or observed energy is

$$\delta U = U' - U$$

$$= \sum_n \sum_{\{j\}} P_{n\{j\}} \left(\frac{\partial E_n}{\partial V}\right)_N \delta V \tag{2.10}$$

From the first law of thermodynamics, which is an expression of conservation of energy, we have for a system undergoing an adiabatic reversible change of volume

$$\delta U = -p \delta V \tag{2.11}$$

Comparison of equations (2.10) and (2.11) gives immediately

$$p = -\sum_n \sum_{\{j\}} P_{n\{j\}} \left(\frac{\partial E_n}{\partial V}\right)_N = -\sum_n \Phi_n \left(\frac{\partial E_n}{\partial V}\right)_N \tag{2.12}$$

where we have used equation (2.3).

It is interesting to note that one can make useful predictions even at this stage, with no knowledge of $P_{n\{j\}}$ apart from the assumnption that it is strongly peaked. For instance the Schrödinger equation for N independent particles in a cube of volume V may readily be solved to show that the E_n are proportional to $V^{-\frac{2}{3}}$. Thus $(\partial E_n/\partial V)_N = -\frac{2}{3}(E_n/V)$ so that $pV = \frac{2}{3}U$ is the predicted equation of state. In general, if E_n is proportional to $V^{-\alpha}$, then $pV = \alpha U$.

Now, returning to our main argument, we see from equation (2.2) that U can also be changed by changing the distribution and keeping the levels fixed, i.e., V and N kept constant. If the final macroscopic state of the system is in thermal equilibrium, then changing the probability distribution for a macroscopic parameter means effectively changing the position of the peak, since, as we have remarked, the only significant features of the distribution function are the position of the peak and the sharpness of the distribution. For the moment we are assuming that this latter point is always valid, so that our statement above must be valid.

If V and N are kept constant and the probability distribution changed, then from equation (2.2)

$$\delta U = \sum_n \sum_{\{j\}} \delta P_{n\{j\}} E_n = \sum_n \delta \Phi_n E_n$$

where $\delta P_{n\{j\}}$ is the change in the probability of the system being in the state $(n, \{j\})$ and $\delta \Phi_n$ is the change in the probability of the system having an energy E_n, i.e.,

$$\delta \Phi_n = \sum_{\{j\}} \delta P_{n\{j\}}$$

Comparison with the first law of thermodynamics for a system at constant volume, i.e., $\delta U = \delta Q$, shows that

$$\delta Q = \sum_n \sum_{\{j\}} \delta P_{n\{j\}} E_n \qquad (2.13)$$

Thus, allowing heat to flow into the system changes the probability distribution but not the energy levels. Now the thermal capacity at constant volume is

$$C_V = \left(\frac{dQ}{dT} \right)_V \qquad (2.14)$$

Hence,

$$C_V = \sum_n \sum_{\{j\}} \frac{dP_{n\{j\}}}{dT} E_n \qquad (2.15)$$

Thus the heat capacity is determined by how the relative probabilities change with temperature (remembering that their sum must always be unity). Now given that an alternative formula for C_V is

$$C_V = T\left(\frac{\partial S}{\partial T}\right)_V \qquad (2.16)$$

we can see that the change in the probabilities is closely related to the change in the entropy. Actually we have jumped ahead because the second formulation of the thermal capacity at constant volume requires that we can write $\delta Q = T\delta S$ and this only follows when we have the second law of thermodynamics. We therefore leave this case at this point and end this chapter with our final choice of boundary conditions for the system.

2.3 Variable number of particles

There is one further case we must consider, and that is the case where the number of particles in the system can vary. Such a system could be one with semi-permeable walls, which would allow molecules to flow in and out of the system from a reservoir. (It should be noted that this case is not so unphysical as may seem at first sight. In fact, in any macroscopic physical system the total number of particles will never be known precisely. In Chapter 4 we will discuss a very physical interpretation of such a system.) To cope with this situation we would have to extend our probability function, to describe the probability of the system having N particles and being in the eigenstate $(n, \{j\})$. We denote this probability by $P_{Nn\{j\}}$. In this case the mean energy of the system can be written

$$U = \sum_{N=0}^{\infty} \sum_{n} \sum_{\{j\}} P_{Nn\{j\}} E_n \qquad (2.17)$$

and the *mean* number $\mathcal{N}$ of particles is

$$\mathcal{N} = \sum_{N=0}^{\infty} \sum_{n} \sum_{\{j\}} P_{Nn\{j\}} N \qquad (2.18)$$

which, from our previous discussion, we shall identify with the observed number.

Now let us suppose that the system is thermally isolated and consider changing the number of particles, keeping the volume constant. The change in the internal energy is

$$\delta U = \sum_{N=0}^{\infty} \sum_{n} \sum_{\{j\}} \delta P_{Nn\{j\}} E_n(N) \qquad (2.19)$$

where $\delta P_{Nn\{j\}}$ is the change in the probability function due to the change in the macroscopic number of particles. Now, in changing the macroscopic number of particles, we have by our previous assumption about the nature of the distribution function, changed the mean value of the distribution, so that

$$\delta P_{Nn\{j\}} = \frac{\partial P_{Nn\{j\}}}{\partial \mathcal{N}} \delta \mathcal{N} \qquad (2.20)$$

Comparison with the first law of thermodynamics for an adiabatic at constant volume, i.e.,

$$\delta U = \mu \delta \mathcal{N} \qquad (2.21)$$

where μ is the chemical potential, gives

$$\mu = \sum_{N=0}^{\infty} \sum_{n} \sum_{\{j\}} \frac{\partial P_{Nn\{j\}}}{\partial \mathcal{N}} E_n(N) \qquad (2.22)$$

This is as far as we can go, without making further deductions about the probability function. At this point it is worth recapitulating the basis on which the results of this section have been made. They rest on one physical law, that of the conservation of energy, and an assumption about the probability function for energy, namely that it is a strongly peaked function and that all values of the energy having a significant probability are macroscopically indistinguishable.

It is this latter assumption that we must now justify. We cannot prove it, any more than one can prove the conservation of energy, which was originally based on experimental evidence, and is now taken to arise from fundamental axioms of physics applying to all branches of natural thermodynamic systems. We shall attempt to justify our statistical assumption by taking a universal law of statistics. This law is assumed to hold for a class of statistical problems which include not only statistical mechanics but also communication theory, economics, etc. The ultimate justification for this law, as for the conservation of energy, must lie in experiment. However, these experiments will encompass a number of branches of human knowledge. The next chapter is therefore devoted to a formulation of this law.

CHAPTER 2: SUMMARY OF MAIN POINTS

1. Application of the law of conservation of energy to an adiabatic reversible change leads to a relationship between the microscopic energy levels and the macroscopic pressure.

2. With a simple assumption about the dependence of the energy levels on volume, an equation of state relating energy, volume and internal energy can be derived.
3. For an irreversible change the probability distribution changes.
4. If the number of particles can change then the chemical potential is determined by how the probabilities change with the mean number of particles.

CHAPTER 3
Statistical Inference

3.1   Introduction
3.2   Uncertainty
3.3   A scale of uncertainty
3.4   Equal probabilities
3.5   Rational requirements for the function H.
3.6   The function H which satisfies conditions A, B, and C
3.7   Unequal probabilities
3.8   Strategy of least bias
3.9   Case 1: toss of a coin
3.10 Case 2: N atoms each with three energy levels
3.11 Case 3: a loaded die
3.12 Case 4: an assembly of $3N$ simple harmonic oscillators
      Questions

3.1   Introduction

As we remarked at the end of the last chapter, we are concerned here with a problem which touches many areas where decisions have to be based on inferences, made against a background of uncertainty or partial ignorance about some relevant factors. This is an everyday affair. For instance, on what grounds does one decide to take a raincoat out on a sunny day? Many factors are involved in determining the pattern of weather at a particular place and time, and one certainly does not know all of them. Consequently one bases a guess on a few relevant facts such as whether it rained yesterday or what the atmospheric pressure reading is. Weather is a nice preliminary example because its patterns are almost infinitely complicated and variable, and prediction is very much a matter of juggling probabilities. We may expect the equilibrium behaviour of solids, liquids and gases to be, by

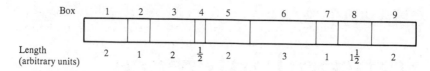

Figure 3.1 A representation of the general problem of uncertainty and prediction. The numbered boxes have different lengths (corresponding to different probabilities with regard to the outcome). The reader is asked to imagine a ball-bearing which travels steadily backwards and forwards above the row of boxes, and which at a randomly specified instant drops into a box.

comparison, *steady*. One reason for this has to do with the high level of prediction of averages which becomes possible where large numbers are involved. For instance, although we may be quite uncertain about the outcome of a single throw of a die, confidence grows with the number of dice until for a large enough number we could be pretty sure that the *average* score would be $3\frac{1}{2}$. This situation can in fact easily be analysed mathematically (see Question 3.1 at the end of this chapter) and an expression obtained for probable deviations from $3\frac{1}{2}$ which may arise for a given number of dice. This is much more predictable than weather, and closer to the kind of problem which typically is amenable to statistical mechanical techniques.

Now the fact is that uncertainty is *reduced* by relevant information. Thus if, while one dithers about the raincoat decision, a friend mentions that it is raining at the moment outside, one's dithering ceases. Information alters decisions by reducing uncertainty. Similarly, if, just before one throws a die, a friend points out that all six sides have the same number 5, one's uncertainty about the outcome is totally removed. So far, so good, but further progress surely must depend on being able to quantify information, to set up a rational scale. In statistical mechanics we are generally concerned with the allocation of particles to particular quantum mechanical states and a notional experiment related to this allocation may be imagined as in Figure 3.1. A row of numbered boxes of equal width but differing lengths (corresponding to different probabilities of occupation) is arranged in a straight line. A ball-bearing travels steadily backwards and forwards above the row and at a randomly specified instant drops into a box. A question it is relevant to consider is this: *before* the ball drops, what is the uncertainty of the outcome? And further, since the uncertainty is obviously zero *after* the ball drops, how much information has been conveyed by the actual result of the experiment?

3.2 Uncertainty

Let us first consider three experiments, each of which has two boxes (outcomes) of different relative lengths (probabilities P_1 and P_2). For example, one experiment could represent tossing a coin, the possible outcomes of which are heads uppermost or tails uppermost, corresponding probabilities being $P_1 = P_2 = 1/2$. In the second different experiment, $P_1 = 3/8$, $P_2 = 5/8$, and in the third, and again different, experiment $P_1 = 1/300$, $P_2 = 299/300$. Now intuitively, we would say that the result of the first experiment is the most uncertain, since we are equally likely to get the first outcome (heads) or the second outcome (tails). On the other hand the third experiment is the least uncertain, since the probability for the second outcome is overwhelmingly large. The uncertainty of the second experiment is intermediate between the other two. If for one of our experiments $P_1 = 0$, $P_2 = 1$, then we would say there is no uncertainty, since we will definitely get the second outcome.

Consider now a different situation, where again we consider three experiments, but this time the possible number of outcomes for each experiment is different, although the probabilities for the possible outcomes of a particular experiment are all equal. Thus, in the first experiment there are two outcomes, $P_1 = P_2 = 1/2$; in the second, four outcomes $P_1 = P_2 = P_3 = 1/4$; and in the third experiment, six outcomes, $P_1 = P_2 = P_3 = P_4 = P_5 = P_6 = 1/6$.

In this set of experiments our intuition would tell us that the last experiment is the most uncertain, simply because there are more outcomes than for the other two and hence the probability of a particular outcome is smaller. The first experiment, on this basis, is the least uncertain. For six racehorses whose forms were identical, the odds given by a bookmaker would be much shorter if only two of the horses competed against each other, than if all six took part in the race!

What we shall endeavour to do now is to find a mathematical measure of a statistical experiment which is consistent with our intuitive concept of uncertainty. It helps a little to think about bent coins, loaded dice, and buckled roulette wheels as a way of thinking about systems which are more likely to be in some states than in some others. The idea is to make predictions about outcomes, just as the idea in statistical mechanics is to make predictions or retrodictions of experimental results.

3.3 A scale of uncertainty

In the light of the considerations above, it is natural to attempt the setting up of a numerical scale of uncertainty by identifying a credible function, H, of probabilities:

$$H = H(P_1, P_2, P_3, \ldots P_j, \ldots)$$

where j ($= 1, 2, 3, \ldots$) labels all the possible outcomes, and P_j is the probability of the jth outcome occurring. The first step is to consider a simple

22   STATISTICAL INFERENCE

case (for example, an unbent coin, an unloaded die, or a well-balanced roulette wheel) in which the P_j can safely be taken as equal.

3.4 Equal probabilities

We consider here the case where the probabilities P_j can be taken as equal. Then

$$P_1 = P_2 = P_3 = \ldots = P_j = \ldots = 1/\Omega$$

where Ω represents the total number of possible outcomes, so that

$$H = H(1/\Omega, 1/\Omega, 1/\Omega, 1/\Omega, \ldots)$$

Since all the P_j are equal in this case, we shall use a shorthand notation by writing this uncertainty as $H(\Omega)$, although strictly we should employ a different symbol.

3.5 Rational requirements for the function H

Firstly, if there is only one possible outcome, then $\Omega = 1$ and there is no uncertainty so that $H(1) = 0$. This conclusion is a *requirement* on the form of H:

If $\Omega = 1$, then $H(\Omega) = 0$ (A)

Secondly, it is also natural to suppose that an experiment for which there are more equally-probable outcomes has more uncertainty than one for which there are fewer:

If $\Omega_1 > \Omega_2$, then $H(\Omega_1) > H(\Omega_2)$ (B)

so that $H(\Omega)$ is a monotonic function of Ω.

Thirdly, we have to consider how to cope with *multiple* events. The simplest case is a double event, for example, the simultaneous throwing of a die (with a number Ω_d of equally-probable outcomes) and spinning of a roulette wheel (with another number Ω_w of equally-probable outcomes). Clearly, the total number of possible outcomes of this multiple experiment is $\Omega_d \Omega_w$, but what of the uncertainty? A rational answer is to be found in the observation that the uncertainty can be reduced to zero in two stages. First, the result of the die can be revealed, followed by the result of the wheel. Or the two revelations can be made in reverse order. In either order, the first revelation alone is insufficient to reduce the uncertainty to zero. We are irresistibly drawn to the conclusion that

If $\Omega = \Omega_1 \Omega_2$, then $H(\Omega) = H(\Omega_1) + H(\Omega_2)$ (C)

3.6 The function H which satisfies conditions A, B, and C

It may seem surprising but there is only one function which will do, and that is the logarithmic function. The proof of this is as follows: we approach it by writing condition (C) in the form

$$H(xy) = H(x) + H(y) \qquad (3.1)$$

where the variables x and y are taken as continuous. This continuity is a small extension of the fact shown above that $H(\Omega)$ is monotonic. In what follows we shall sometimes use the notation (LHS) for $H(xy)$, and (RHS) for $H(x) + H(y)$. We use the device of differentiating both sides of the equation firstly with respect to x (keeping y constant) and secondly with respect to y (keeping x constant). Thus,

$$\left[\frac{\partial(\text{LHS})}{\partial x}\right]_y \equiv \left[\frac{\partial H(xy)}{\partial x}\right]_y \qquad (3.2)$$

$$= \frac{dH(xy)}{d(xy)} \left[\frac{\partial(xy)}{\partial x}\right]_y \qquad (3.3)$$

$$= \frac{dH(xy)}{d(xy)} y \qquad (3.4)$$

while

$$\left[\frac{\partial(\text{RHS})}{\partial x}\right]_y = \frac{dH(x)}{dx} \qquad (3.5)$$

So

$$\frac{dH(xy)}{d(xy)} y = \frac{dH(x)}{dx} \qquad (3.6)$$

Similarly,

$$\frac{dH(xy)}{d(xy)} x = \frac{dH(y)}{dy} \qquad (3.7)$$

So, from (3.6) and (3.7),

$$xy \frac{dH(xy)}{d(xy)} = x \frac{dH(x)}{dx} = y \frac{dH(y)}{dy} \qquad (3.8)$$

Now the second part of (3.8) depends only on x while the third depends only on y, and since x and y can be independently varied, the three parts of (3.8) can only be equal to a constant. That is,

24   STATISTICAL INFERENCE

$$x\frac{dH(x)}{dx} = y\frac{dH(y)}{dy} = C_0 \tag{3.9}$$

where C_0 is a constant. Integrating,

$$H(x) = C_0 \ln x + C_1 \tag{3.10}$$

The integration constant C_1 must be zero for (3.10) to satisfy (A), and we are obviously at liberty to choose the scale constant C_0 to be unity. Therefore the required form for the case of equal probabilities is

$$H(\Omega) = \ln \Omega \tag{3.11}$$

3.7 Unequal probabilities

Consider for concreteness a buckled roulette wheel, the probabilities P_j of whose outcomes j are unequal. We want a form for H, and we adopt the mental device of imagining the wheel to be spun a very large number N of times; in fact we shall at the end of the argument take N to the limiting value of infinity. In these imagined circumstances, each outcome j would occur NP_j times, and these would be *numbers of occurrences* rather than merely *probabilities*. The multiple experiment, consisting of N spins, must have NP_1 outcomes of type 1, NP_2 of type 2, NP_3 of type 3, etc., but these may occur in many different orders. Thus the original uncertainty about *which* outcome has by a mental device been converted into an uncertainty about *order*. And this is easy to solve, since all the possible orders which may occur in an N-spin experiment are equally likely so that can use equation (3.11) for the associated uncertainty H_N:

$$H_N = \ln \Omega = \ln \left(\frac{N!}{\prod_j (NP_j)!} \right) \tag{3.12}$$

where $\prod (NP_j)!$ represents the product $(NP_1)! \times (NP_2)! \times (NP_3)! \times \ldots$ all the factorials of the numbers NP_j, and $N!/\prod(NP_j)!$ is the number of equally-probable ways of reordering the sequence of outcomes. The final step is to use requirement (B) above to deduce that

$$H_N = NH \tag{3.13}$$

where H is the uncertainty associated with just one spin of the wheel. Hence

$$H = \frac{1}{N}\ln\left(\frac{N!}{\prod_j (NP_j)!}\right) \qquad (3.14)$$

All that is required now to reduce this to a manageable form is to let N approach infinity so that Stirling's approximation $\ln X! \simeq X\ln X - X$ can be used for all the large numbers $X = N$ and $X = NP_j$. The final result is

$$H = -\sum_j P_j \ln P_j \qquad (3.15)$$

This is the general function we have sought to identify. It reduces, as it must, to the special case of equation (3.11) when $P_j = 1/\Omega$ for all j. Moreover it is always positive since all the P_j are in the range 0 to 1.

3.8 Strategy of least bias

What we have shown is that if the P_j are known, then H can be calculated from equation (3.15). In statistical mechanics however, as in roulette, the problem is usually the other way around in that the P_j are precisely the quantities which we *want* to know. We may, or may not, be given certain information about the experiment, for example, one or more mean values. Clearly without further information we cannot determine these probabilities uniquely. There will be a large number of probability distributions that will be consistent with the given information. What seems reasonable therefore is to choose a distribution that is consistent with the given information but contains no particular bias. This is similar to the situation when choosing samples in statistical surveys. For example, if you were trying to determine the age distribution of a country, you would not choose your sample to consist solely of old peoples' homes! To return to the problem of choosing the probabilities, what we are saying is simply that we must not give a significantly different probability to any particular outcome unless the given information justifies it. The obvious way of achieving this is to maximize the uncertainty subject to the given information. This will of course lead to a probability distribution which is not unique, and we can only verify its appropriateness by comparison with experiment. It does, however, give us a specified procedure which we can use consistently. We now consider some examples of this procedure; further examples are set as problems at the end of the chapter.

3.9 Case 1: toss of a coin

We *know* that for a properly balanced coin, the probabilities P_h(head) and P_t(tail) are both equal to $\frac{1}{2}$ - but let us *prove* it using the strategy of least bias.

$$H = -\sum_j P_j \ln P_j = -(P_h \ln P_h + P_t \ln P_t)$$
$$= -(P_h \ln P_h + (1 - P_h) \ln(1 - P_h)) \qquad (3.16)$$

because of the general constraint that $(P_h + P_t) = 1$. Now expression (3.16) is a simple function of one variable P_h and can readily be maximized by differentiation:

$$\frac{dH}{dP_h} = \ln\left(\frac{1 - P_h}{P_h}\right) = 0 \text{ for } P_h = \tfrac{1}{2} \qquad (3.17)$$

and

$$\frac{d^2H}{dP_h^2} = -\left(\frac{1}{P_h} + \frac{1}{1 - P_h}\right) \qquad (3.18)$$

which is negative for $P_h = \tfrac{1}{2}$ (or indeed for any value of P_h in the allowed range 0 to 1) so that H indeed has its maximum value for $P_h = P_t$ as expected.

3.10 Case 2: N atoms each with three energy levels

We take as given that the three levels have 0, 1, and 2 units (with each atom having corresponding unknown probabilities P_0, P_1, and P_2 of being in those levels) and that the total energy is 0.3 N units, and set out to calculate P_0, P_1, and P_2. This example is more obviously related to statistical mechanics than the simple toss of a coin, but the mathematical approach is very similar.

$$H = -\sum_j P_j \ln P_j = -(P_0 \ln P_0 + P_1 \ln P_1 + P_2 \ln P_2) \qquad (3.19)$$

$$= -(P_0 \ln P_0 + (1.7 - 2P_0) \ln(1.7 - 2P_0)$$
$$+ (P_0 - 0.7) \ln(P_0 - 0.7)) \qquad (3.20)$$

where the second line follows by use of the general constraint $(P_0 + P_1 + P_2) = 1$ and the energy constraint, $((0 \times P_0) + (1 \times P_1) + (2 \times P_2)) = P_1 + 2P_2 = 0.3$, to eliminate P_1 and P_2. Since only one variable, P_0, is then involved we can again proceed by simple differentiation to obtain

$$\frac{dH}{dP_0} = \ln\left(\frac{(1.7 - 2P_0)^2}{P_0(P_0 - 0.7)}\right) \qquad (3.21)$$

For this expression to be zero, the argument of the logarithm must be unity. Hence, we have a quadratic equation which can readily be solved for P_0. The values of P_1 and P_2 follow from going back to the two constraints. The results are

$$\left.\begin{aligned}P_0 &= (6.1 - (2.53)^{\frac{1}{2}})/6 \simeq 0.752 \\ P_1 &= ((2.53)^{\frac{1}{2}} - 1)/3 \simeq 0.197 \\ P_2 &= (1.9 - (2.53)^{\frac{1}{2}})/6 \simeq 0.051\end{aligned}\right\} \qquad (3.22)$$

3.11 Case 3: a loaded die

The *unloaded* die is merely a special case, so we prefer to tackle the more general problem. As we shall see, a new mathematical technique is required. (Appendix 1 describes this technique in general terms.) We have to maximize

$$H = -\sum_{j=1}^{6} P_j \ln P_j \qquad (3.23)$$

subject to the general constraint

$$\sum_{j=1}^{6} P_j = 1 \qquad (3.24)$$

and an extra given constraint relating to the observed average score which we shall denote by A. An unloaded die would have $A = 3\frac{1}{2}$, but a loaded die may have any value in the range 1 to 6, depending on the character of the loading. This extra constraint can be written as

$$\sum_{j=1}^{6} jP_j = A \qquad (3.25)$$

It is easy to see that H cannot be written in terms of one variable as in cases 1 and 2 above. How is it to be maximized? We imagine ourselves to be in the neighbourhood of the maximum with the probabilities P_j close to their optimum values but still slightly adjustable; we shall use an asterisk in P_j^* and H^* as a reminder of this adjustability. We seek to maximize H by setting

28 STATISTICAL INFERENCE

$$dH^* = -\sum_{j=1}^{6} d(P_j^* \ln P_j^*) = -\sum_{j=1}^{6} (1 + \ln P_j^*) dP_j^* = 0 \qquad (3.26)$$

Also, for *any* set of P_j including the P_j^*, their sum must be unity, so

$$\sum_{j=1}^{6} dP_j^* = 0 \qquad \text{(general constraint)} \qquad (3.27)$$

and similarly

$$\sum_{j=1}^{6} j\, dP_j^* = 0 \qquad \text{(extra constraint)} \qquad (3.28)$$

Now we introduce two parameters λ and β as 'undetermined multipliers' which can take any values in the linear combination containing the three independently zero sums (3.26), (3.27), and (3.28):

$$\sum_{j=1}^{6} (1 + \ln P_j^*) dP_j^* + \lambda \sum_{j=1}^{6} dP_j^* + \beta \sum_{j=1}^{6} j\, dP_j^* = 0 \qquad (3.29)$$

or

$$\sum_{j=1}^{6} (1 + \ln P_j^* + \lambda + \beta j) dP_j^* = 0 \qquad (3.30)$$

Now λ and β may take *any* values so we are free to choose them as we wish. We *decide* to choose λ and β in such a way as to make the first two brackets, i.e., those with $j = 1$ and $j = 2$ in the sum, independently zero. This means setting

$$\left. \begin{array}{l} \lambda = \ln P_2^* - 2\ln P_1^* - 1 \\ \text{and} \quad \beta = \ln P_1^* - \ln P_2^* \end{array} \right\} \qquad (3.31)$$

Since we do not yet know P_1^* or P_2^* it may seem that little progress has been made. The point is, however, that having set the first two brackets in the sum to be zero, we have made sure that the four remaining terms are also independently zero since dP_3^*, dP_4^*, dP_5^*, and dP_6^* can be freely varied, the two constraints having accounted for the values of P_1^* and P_2^*. We conclude that all six brackets in the sum (3.30) are independently zero so that

$$\ln P_j^* = -1 - \lambda - \beta j \qquad (3.32)$$

that is,

CASE 4: AN ASSEMBLY OF 3N SIMPLE HARMONIC OSCILLATORS

$$P_j^* = \exp(-1 - \lambda)\exp(-\beta j) \tag{3.33}$$

Now we know that the sum of all the P_j^* is unity, so λ can be eliminated and we are left with

$$P_j = \frac{\exp(-\beta j)}{\sum_j \exp(-\beta j)} \tag{3.34}$$

The asterisk is no longer necessary and has been dropped at this point, since the adjustments needed to locate the maximum in H have now been made. The final step is to go back to the constraint (3.28) that the average be A and this leads to an equation for β, though admittedly in a form which can only be solved numerically:

$$A = \frac{\exp(-\beta) + 2\exp(-2\beta) + 3\exp(-3\beta) + 4\exp(-4\beta) + 5\exp(-5\beta) + 6\exp(-6\beta)}{\exp(-\beta) + \exp(-2\beta) + \exp(-3\beta) + \exp(-4\beta) + \exp(-5\beta) + \exp(-6\beta)} \tag{3.35}$$

The sums in the numerator and denominator of equation (3.35) can be expressed in neater forms:

$$A = \frac{1 - 7\exp(-6\beta) + 6\exp(-7\beta)}{(1 - \exp(-\beta))(1 - \exp(-6\beta))} \tag{3.36}$$

Some special cases can be seen by careful inspection. For example for $A = 3\frac{1}{2}$ (unloaded die), then $\beta = 0$ and the six probabilities are equal to 1/6, the commonsense result. If $A = 1$ or $A = 6$ (extreme cases) then $\beta = +\infty$ or $\beta = -\infty$ respectively. If $A = 4$, a non-special case, then iteration using a programmable pocket calculator will deliver the results (correct to four places of decimals) that $\beta = -0.1746$ so that $P_1 = 0.1031$, $P_2 = 0.1227$, $P_3 = 0.1461$, $P_4 = 0.1740$, $P_5 = 0.2072$, and $P_6 = 0.2469$. One can also find λ by going back to equation (3.33) and the value is $\lambda = 1.4470$.

3.12 Case 4: an assembly of 3N simple harmonic oscillators

This is a very physical situation, but also one which logically relates closely to the cases above. We imagine $3N$ oscillators, each of which may be (according to quantum mechanics) in one of the energy levels $\epsilon = h\nu(j + \frac{1}{2})$ where $h\nu$ is a constant and j may take any integral value including zero. The question we ask is: what are the probabilities of a given oscillator being in particular levels if the average energy is taken to be u? We have to maximize

$$H = -\sum_{j=0}^{\infty} P_j \ln P_j \tag{3.37}$$

subject to the general constraint

$$\sum_{j=0}^{\infty} P_j = 1 \tag{3.38}$$

and an extra constraint which can be written as

$$\sum_{j=0}^{\infty} h\nu(j + \tfrac{1}{2})P_j \equiv h\nu \sum_{j=0}^{\infty} jP_j + \tfrac{1}{2} h\nu = u \tag{3.39}$$

making use of (3.38). On slight rearrangement this becomes

$$\sum_{j=0}^{\infty} jP_j = (u/h\nu) - \tfrac{1}{2} \tag{3.40}$$

The reader should now slightly modify the algebra following the very similar equations (3.23), (3.24), and (3.25) to obtain the identical equation to (3.34),

$$P_j = \frac{\exp(-j\beta)}{\sum_{j=0}^{\infty} \exp(-j\beta)} \tag{3.41}$$

and that corresponding to (3.35):

$$(u/h\nu) - \tfrac{1}{2} = \frac{\sum_{j=0}^{\infty} j \exp(-j\beta)}{\sum_{j=0}^{\infty} \exp(-j\beta)} \tag{3.42}$$

The infinite sums can be performed explicitly since the denominator is an infinite geometric series and the numerator can be derived from the denominator by differentiation with respect to β, and this gives a satisfying simplification corresponding to (3.36):

$$(u/h\nu) - \tfrac{1}{2} = [\exp(+\beta) - 1]^{-1}$$

which on rearrangement gives

$$\beta = \ln\left[\frac{(u/h\nu) + \frac{1}{2}}{(u/h\nu) - \frac{1}{2}}\right] \tag{3.43}$$

Substitution back into (3.41) and rearrangement now gives P_j and, by simple extension, the expected number of oscillators $3NP_j$ in level j:

$$3NP_j = 3N\frac{[(u/h\nu) - \frac{1}{2}]^j}{[(u/h\nu) + \frac{1}{2}]^{j+1}} \tag{3.44}$$

The object of this chapter has been to lay down a procedure for determining the probability of an outcome, consistent with the given information. This procedure can be stated as a law:

> 'The most probable distribution is given by maximizing the uncertainty, subject to the constraints imposed by the given information.'

We have given a few examples of the operation of the procedure, and a few more can be found in the problem section. Whether this law is correct cannot be proved but only verified by experiment. This is equally true of the law of conservation of energy, although in the present case the law can be tested not only in physics but in several other subjects. We shall take it that this has been done and assume its validity, as we have for the law of conservation of energy.

CHAPTER 3: SUMMARY OF MAIN POINTS

1. The related notions of information and uncertainty are introduced in the context of common situations in which the prediction of the outcome of an event is a matter of probabilities rather than certainties.
2. By a series of logical steps, it is shown that uncertainty can be quantified on a rational scale.
3. Uncertainty must be a function of the probabilities of the various possible outcomes. Where these can be taken as equal, for example in the throwing of dice, it is shown that it is rational to associate uncertainty H with $\ln\Omega$ where Ω is the number of possible outcomes (6 in the case of a die).
4. Further, it is shown that where the probabilities cannot be assumed to be equal, for example in a loaded roulette wheel, the more general function H of probabilities P_j is $H = -\Sigma P_j \ln P_j$.
5. If, as is usual in applications of these ideas to physical situations, the P_j are not known in advance then it is argued that the best way of assigning them is to require that the uncertainty H be maximized.
6. This enables the P_j to be uniquely deduced using, where necessary, the method of Lagrange multipliers.

32   STATISTICAL INFERENCE

7. This strategy of least bias is applied to a number of situations ranging from dice throws to assemblies of quantum simple harmonic oscillators.

QUESTIONS

3.1 Consider the experiment of simultaneously throwing a number N of dice, a process referred to in Section 3.1. The task is to calculate the expectation values of the total score X and that of the r.m.s. deviation defined as

$$\langle \Delta X \rangle_{\text{rms}} = (\langle X^2 \rangle - \langle X \rangle^2)^{\frac{1}{2}}$$

To do this it is only necessary to find $\langle X \rangle$ and $\langle X^2 \rangle$. Use the following notation: let the probability that the αth die scores $x_{i\alpha}$ be $p_{i\alpha}$. Clearly the possible values of x_i (i.e. $x_{i\alpha}$) are 1, 2, 3, 4, 5 or 6, and assuming all the dice to be well behaved, $p_{i\alpha} = p_i = 1/6$.

(a) Prove that $\displaystyle\langle X \rangle = \sum_{i=1}^{6} \sum_{\alpha=1}^{N} x_{i\alpha} P_{i\alpha} = \sum_{\alpha=1}^{N} \langle x_{i\alpha} \rangle = \frac{7N}{2}$

(b) Prove that $\displaystyle\langle X^2 \rangle = \sum_{\alpha=1}^{N} \sum_{\beta=1}^{N} \langle x_{i\alpha} x_{i\beta} \rangle$

$$= \sum_{\alpha=1}^{N} \langle x_{i\alpha}^2 \rangle + \sum_{\alpha \neq \beta}^{N} \langle x_{i\alpha} \rangle \langle x_{i\beta} \rangle$$

$$= N \langle x_i^2 \rangle + N(N-1)(\langle x_i \rangle)^2$$

(c) Hence prove that $\langle \Delta X \rangle_{\text{rms}} = N(\langle x_i^2 \rangle - (\langle x_i \rangle)^2)$

and

$$\frac{\langle \Delta X \rangle^2_{\text{rms}}}{\langle X \rangle} = \left(\frac{20}{1029N} \right)^{\frac{1}{2}}$$

The point to observe is that the sharpness of the distribution increases with increasing N.

3.2 Associated with each of the following sets of probabilities is an uncertainty H. Calculate H for each case and establish which has the largest value.

(a) $\frac{1}{3}, \frac{1}{3}, \frac{1}{3}$

(b) $\frac{1}{2}, \frac{1}{3}, \frac{1}{6}$

(c) $\frac{1}{4}, \frac{1}{4}, \frac{1}{4}, \frac{1}{4}$

(d) $\frac{3}{4}, \frac{1}{8}, \frac{1}{8}, 0$

3.3 Can you find any objections to the use of the following alternative functional forms for the uncertainty?

(a) $$H = \sum_j P_j^2$$

(b) $$H = \sum_j \ln P_j$$

3.4 For each of the following cases calculate (a) the probability of obtaining the result '1' in a single experiment, and (b) the uncertainty with regard to the outcome before the experiment is performed.

(i) The toss of a coin (head = 1, tail = 2).

(ii) The throw of a die.

(iii) The spin of a roulette wheel in which there are 27 equally sized sections in the wheel, four of which are identically labelled '1' and the 23 others all have distinct labels.

3.5 In an examination it is possible to have marks deducted as well as credited and the possible marks for a particular question are 1, 0, or −1. If you only know that the mean square mark for a whole group of students is 0.5, what should be your best estimates of the probabilities of a particular student obtaining marks 1, 0, or −1? (*As in Question 3.6 below, it is not necessary to use the method of Lagrangian multipliers and it is in fact easier not to. You might learn something, however, by finding two ways of solving this problem, one with and the other without the use of Lagrangian multipliers.*)

3.6 (a) For a toss of a coin, the uncertainty H with regard to the outcome (h = head; t = tail) is

$$H = -\sum_j P_j \ln P_j = -P_h \ln P_h - P_t \ln P_t$$

$$= -P_h \ln P_h - (1 - P_h) \ln (1 - P_h)$$

where use is made of the fact that $P_h + P_t = 1$. Show that this uncertainty has a maximum value of ln2 for $P_h = P_t = \tfrac{1}{2}$.

(b) This is mathematically similar to part (a) but is closer to physics. An insulated vessel contains 1000 molecules of a pure gas. From quantum theory and spectroscopic measurements it is known that at any instant of time a given molecule may possess 0, 1, or 2 units of energy. Estimate the most probable number of molecules in each of levels 0, 1, and 2. (*As in Question 3.5 above, it is not necessary to use the method of Lagrangian multipliers and it is in fact easier not to. You might learn something, however, by finding two ways of solving this problem, one with and the other without the use of Langrangian multipliers.*)

3.7 In Chapter 4 we shall discuss the 'canonical distribution'. The analysis is very similar to that of a loaded die leading to equation (3.34). Following that analysis prove that, for a system whose allowed states are labelled j with energies E_j and whose average energy is given to be U, the probability P_j of finding the system in state j is given by

$$P_j = \frac{\exp(-\beta E_j)}{Q}$$

where

$$Q = \sum_j \exp(-\beta E_j)$$

[The mathematical task is to maximize $H = -\Sigma P_j \ln P_j$ (corresponding to equation (3.23)) subject to the constraints $\Sigma P_j = 1$ (corresponding to equation (3.24)) and $\Sigma E_j P_j = U$ (corresponding to equation (3.25)).]

CHAPTER 4
The Law of Maximum Uncertainty

4.1 Introduction
4.2 Thermodynamically isolated system (microcanonical distribution)
4.3 System in a heat bath (canonical distribution)
4.4 System in a heat-and-particle bath (grand canonical distribution)
Questions

4.1 Introduction

From our discussion in the previous chapter, it is clear that the first step is to decide what information we have or explicitly assume about the system. There are several possibilities, depending on the system and on the specific boundary conditions. We shall consider three types of boundary condition: a thermodynamically isolated system, a system of a heat bath, and a system in a heat-and-particle bath.

4.2 Thermodynamically isolated system (microcanonical distribution)

Let us consider first a thermodynamically isolated system whose macroscopic energy U we have measured. If ΔU is the smallest energy difference that is measurable macroscopically, then on a microscopic scale we can say that our system will have an actual energy eigenvalue E_n somewhere in the range given approximately by $(U - \tfrac{1}{2}\Delta U) < E_n < (U + \tfrac{1}{2}\Delta U)$. Thus the accessible states of the system must have energies in this range, and we can write

$$P_{n\{j\}} = 0 \text{ if } E_n < (U - \tfrac{1}{2}\Delta U), \text{ or if } E_n > (U + \tfrac{1}{2}\Delta U) \tag{4.1}$$

Apart from defining the accessible states of the system we have no other information, so that we maximize the uncertainty (defined by equation 3.15)

$$H = -\sum_n{}' \sum_{\{j\}}{}' P_{n\{j\}} \ln P_{n\{j\}} \tag{4.2}$$

where the primes in the sums indicate that we are to sum only over those states that satisfy the condition $(U - \tfrac{1}{2}\Delta U) < E_n < (U + \tfrac{1}{2}\Delta U)$. As in the case of an unloaded die referred to in the lines following equation (3.36), this leads to equal probabilities for states, i.e.

$$P_{n\{j\}} = \frac{1}{\Omega(U, \Delta U, V)} \text{ for } (U - \tfrac{1}{2}\Delta U) < E_n < (U + \tfrac{1}{2}\Delta U) \tag{4.3}$$

where $\Omega(U, \Delta U, V)$ is the number of states in the range $(U - \tfrac{1}{2}\Delta U)$ to $(U + \tfrac{1}{2}\Delta U)$, and V is the volume of the system. The other states will of course have probabilities equal to zero, since by the definition they are not possible states of the system consistent with the given information. The corresponding value of the maximum uncertainty (see equation 3.11) is

$$H_m = \ln \Omega(U, \Delta U, V) \tag{4.4}$$

Now let us consider connecting our system to a heat bath and allowing a small amount of heat δQ to flow adiabatically slowly into the system. We then isolate the system again. Now we know from our arguments in Chapter 2 that the heat flowing into the system changes the probability functions and hence the macroscopic energy U.

For constant volume and ΔU, equation (4.4) gives

$$\delta H_m = \frac{1}{\Omega} \left(\frac{\partial \Omega}{\partial U} \right)_{V, \Delta U} \delta U \tag{4.5}$$

or

$$\delta U = \frac{\Omega \delta H_m}{\left(\dfrac{\partial \Omega}{\partial U} \right)_{V, \Delta U}} \tag{4.6}$$

$$= \frac{\delta H_m}{\left(\dfrac{\partial \ln \Omega}{\partial U} \right)_{V, \Delta u}} \tag{4.7}$$

THERMODYNAMICALLY ISOLATED SYSTEM

Up until now we have used the physical law of conservation of energy as expressed in the first law of thermodynamics and the statistical law of maximum uncertainty. To make further progress we have to make use of a second physical law, namely, the second law of thermodynamics. As commonly formulated in terms of the efficiency of a Carnot engine it is not easy to see how to use it. So rather than use the law itself, we will make use of one of its consequences. The second law uses the concept of entropy to write $\delta Q = T \delta S$ and hence the thermodynamic identity

$$T\delta S = \delta U + p\delta V \tag{4.8}$$

If we compare this expression at constant volume with (4.7) we get

$$\delta S = \frac{\delta H_m}{T \left(\frac{\partial \ln \Omega}{\partial U} \right)_{V, \Delta U}} \tag{4.9}$$

Now since, in the infinitesimal limit, dS is a perfect differential, it follows that

$$T \left(\frac{\partial \ln \Omega}{\partial U} \right)_{V, \Delta U}$$

must be a function of H_m say $f(H_m)$ so that

$$S = \int \frac{dH_m}{f(H_m)} + S_0 \tag{4.10}$$

or

$$S = Q(H_m) + S_0 \tag{4.11}$$

where S_0 is a constant which, according to the third law of thermodynamics, may be taken to be zero.

To identify the function $Q(H_m)$ we make use of the fact that both S and H_m are additive, so that, for two systems A and B,

$$S_{AB} = S_A + S_B \text{ or } Q(H_A + H_B) = Q(H_A) + Q(H_B) \tag{4.12}$$

(for clarity we drop the subscript m, and from now on H is to be interpreted as the maximum uncertainty given by equation (4.4)). Differentiating both sides of this last equation with respect to H_A and H_B gives

$$\frac{dQ(H_A + H_B)}{d(H_A + H_B)} = \frac{dQ(H_A)}{dH_A} = \frac{dQ(H_B)}{dH_B} = k \tag{4.13}$$

where k is a constant, so that

$$Q(H) = kH$$

and from equation (4.11) with $S_0 = 0$,

$$S = kH = k\ln\Omega(U, \Delta U, V) \qquad (4.14)$$

Furthermore, since from equation (4.8)

$$\frac{1}{T} = \left(\frac{\partial S}{\partial U}\right)_{V,\Delta U} \qquad (4.15)$$

we have

$$kT = \frac{\Omega}{\left(\frac{\partial \Omega}{\partial U}\right)_{V,\Delta U}} = \frac{1}{\left(\frac{\partial \ln\Omega}{\partial U}\right)_{V,\Delta U}} \qquad (4.16)$$

which is consistent with equations (4.8) and (4.14). We shall postpone for the moment the identification of the constant k.

Equation (4.14) is extremely interesting in identifying, apart from a scaling constant, the uncertainty with the entropy of the system. Thus entropy is not, strictly speaking, a physical quantity in the same way as energy. A body does not possess entropy – to quote Jaynes (*Physics Review*, **106**, p. 620 (1957)): 'Entropy is an anthropomorphic concept. For it is a property, not of the physical system but of the particular experiments you or I choose to perform on it.' This is easily illustrated from the expression for S given by equation (4.14). In principle, one could increase the accuracy of the device used for measuring the energy, and hence reduce the range ΔU. In that case $\Omega(U, \Delta U, V)$ will be reduced and hence the entropy. In practice, any instrumental improvement resulting in a reduction in ΔU by factor M will cause a subtraction from S or $k\ln M$ which will always be negligible when compared with $k\ln\Omega$. Only if we had a complete microscopic description of the state of the system, i.e., we know the quantum numbers associated with every state $(n, \{j\})$, is the entropy unambiguous, i.e., it is zero. Then, however, our macroscopic notion of entropy is not defined. Now our intention as stated was to prove, from the principle of maximum uncertainty, that the relevant distribution functions had the properties that they were strongly peaked and that all values within that peak were macroscopically identical. However, for the systems considered in this section, the probability function for energy has by definition these properties. Strictly one should show the coefficient of variation for any other macroscopic variable is vanishingly small, but this will not be attempted here.

Having, we hope, thrown some light on the nature of entropy, we can now turn to the more practical task of calculating other thermodynamic quantities. These are easily found from standard thermodynamic relations; the pressure

$$p = T \left(\frac{\partial S}{\partial V}\right)_{U,\Delta V} = kT \left(\frac{\partial \ln \Omega}{\partial V}\right)_{U,\Delta V} \quad (4.17)$$

Since Ω is a function of U and V, the expressions (4.14) for S and (4.17) for p are both functions of U and V. From the point of view of thermodynamics we require these variables as functions of T and V. To do this we invert (4.16) and find U as a function of T and V. Substitution in (4.14) and (4.17) will then give S and p as functions of the normal independent variables. Once this has been achieved other thermodynamic quantities can be found, for example, the heat capacity at constant

$$C_V = \left(\frac{\partial U}{\partial T}\right)_V \quad (4.18)$$

and the isothermal bulk modulus

$$B = -V \left(\frac{\partial p}{\partial V}\right)_T \quad (4.19)$$

Finally we have to identify the constant k. It can be shown that for a Maxwell–Boltzmann gas discussed in Chapter 9, $\Omega(U, \Delta U, V)$ is given by

$$\Omega(U, \Delta U, V) = C U^{(3N/2)} V^N \Delta U \quad (4.20)$$

where N is the number of particles and C is a constant independent of U and V. Hence

$$\ln \Omega = (3N/2) \ln U + N \ln V + \ln(C \Delta U) \quad (4.21)$$

We are now in a position to calculate the macroscopic parameters for a Maxwell–Boltzmann gas. From equation (4.16)

$$kT = \frac{2U}{3N} \quad (4.22)$$

or

$$U = \frac{3NkT}{2} \quad (4.23)$$

and from equation (4.17)

$$p = \frac{kTN}{V} \tag{4.24}$$

From equation (4.23) or (4.24) we see that k must be identified with Boltzmann's constant ($k = 1.38 \times 10^{-23} J/K$).

In practice, the microcanonical distribution is not very easy to use. The problem is that one has to evaluate the number of states subject to the restriction that the total energy lies in a certain range, and only in a few cases can this be done exactly, one of which is the so-called Einstein crystal. This is discussed at greater length in Chapter 6, where the calculation is done using the canonical distribution (see the following section). It is done here using the microcanonical distribution as an example of how, in nearly all cases, the different distributions lead to the same result. The exceptions to this rule occur at critical points (see later in this chapter for a fuller discussion). Essentially the solid is modelled as a collection of $3N$ oscillators of frequency ω. Question 4.11 at the end of this chapter takes the reader through the problem of calculating $\Omega(U, \Delta U, V)$. The result is

$$\ln[\Omega(U, \Delta U, V)] = [U/\hbar\omega + \tfrac{3}{2}N]\ln[U/\hbar\omega + \tfrac{3}{2}N]$$
$$- [U/\hbar\omega - \tfrac{3}{2}N]\ln[U/\hbar\omega - \tfrac{3}{2}N]$$

Using equation (4.16) we find

$$\frac{1}{kT} = \frac{1}{\hbar\omega}\ln\left[\frac{U + \tfrac{3}{2}N\hbar\omega}{U - \tfrac{3}{2}N\hbar\omega}\right]$$

which may be solved for U to give

$$U = \tfrac{3}{2}N\hbar\omega \coth(\hbar\omega/2kT)$$

Discussion of this result is postponed to Chapter 6.

4.3 System in a heat bath (canonical distribution)

The next type of system we consider, in increasing order of physical complexity, is one which is immersed in a heat bath. The heat bath is assumed so large that any energy transferred from the system to the heat bath is negligible compared to the total energy of the heat bath. This is shown in Figure 4.1.

The total energy U_{tot} of the system plus heat bath will be

$$U_{tot} = U + U_{hb} + U_{int} \tag{4.25}$$

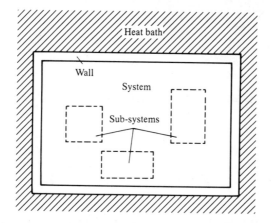

Figure 4.1 System in contact with an infinite heat bath through walls which transmit heat but no particles, corresponding to conditions for use of the canonical distribution. (If the walls were impenetrable to heat *and* particles, the microcanonical distribution would be appropriate; if the walls allowed passage to heat *and* particles, the grand canonical distribution would be appropriate.) The dashed lines enclose macroscopic sub-systems of a system as discussed in the text.

where U is the energy of the system, U_{hb} that of the heat bath, and U_{int} the interaction energy, assumed to be negligible. Now since U_{tot} has to be constant, U can range from the ground state energy to energies approaching U_{tot} which, because of our definition of the heat bath, tends to infinity. Consequently all states of the system are available, and, if this was all the information we had, it follows from our discussion of the previous section that the entropy of the system would be the logarithm of the total number of states, which would be infinite! In such a situation, however, the system would not be in a well-defined thermodynamic state. Assuming that the volume is known, we need one other thermodynamic variable to define the macroscopic state of the system. It is convenient to choose the observed energy for this parameter, which according to our previous discussion we can identify with the mean value over the probability distribution.

Thus, for the system in a heat bath, we have in principle a non-zero probability of the system being any of the eigenstates, unlike the isolated system where the probability was zero for all states with energies outside a well-defined energy range. In practice, for the system in a heat bath, we hope in accordance with our previous discussion that the probabilities will be strongly peaked about the observed values. We must now attempt to find the most likely probability $P_{n\{j\}}$ of the system being in state $(n, \{j\})$, given that all states are accessible and that the mean energy is U (the observed value). According to the discussion in Chapter 3 we maximize the uncertainty

$$H = -\sum_n \sum_{\{j\}} P_{n\{j\}} \ln P_{n\{j\}} \qquad (4.26)$$

subject to constraints

$$\sum_n \sum_{\{j\}} P_{n\{j\}} = 1 \qquad (4.27)$$

and

$$\sum_n \sum_{\{j\}} P_{n\{j\}} E_n = U \qquad (4.28)$$

This is just like the problems we discussed in Section 3.2, and was the subject of Question 3.7. We find

$$P_{n\{j\}} = \frac{\exp(-\beta E_n)}{Q} \qquad (4.29)$$

and for the corresponding maximum uncertainty

$$H_m = \ln Q + \beta U \qquad (4.30)$$

where

$$Q = \sum_n \sum_{\{j\}} \exp(-\beta E_n)$$
$$= \sum_n g_n \exp(-\beta E_n) \qquad (4.31)$$

and g_n is the degeneracy of the energy level E_n, i.e., the number of states having energy E_n. The probability distribution is known as the Gibbs canonical distribution and Q as the canonical partition function.

Now, as in the case of the isolated system discussed in the previous section, let us allow a small amount of heat δQ to flow reversibly into our system keeping the volume constant. In practice this is achieved by altering the temperature of the heat bath. From the arguments in Chapter 2 this will produce a change in the probability distribution and hence a change in the maximum uncertainty

$$\delta H_m = -\sum_n \sum_{\{j\}} \delta(P_{n\{j\}} \ln P_{n\{j\}})$$
$$= +\beta \sum_n \sum_{\{j\}} \delta P_{n\{j\}} E_n = \beta \delta U \qquad (4.32)$$

where the $P_{n\{j\}}$ are given by equation (4.29) and we have used the fact that

$$\sum_n \sum_{\{j\}} \delta P_{n\{j\}} = 0$$

Comparison with the thermodynamic identity at constant volume, $\delta U = T\delta S$, gives

$$\delta S = \frac{\delta H_m}{\beta T} \tag{4.33}$$

Now since δH_m and δS are both perfect differentials, βT must be a function of H_m. Equation (4.33) is thus analogous to equation (4.9), and similar arguments which led to equation (4.14) yield

$$S = kH_m \tag{4.34}$$

and

$$\beta = \frac{1}{kT} \tag{4.35}$$

Hence using equation (4.30),

$$S = k \ln Q + \frac{U}{T} \tag{4.36}$$

or introducing the Helmholtz free energy

$$F = U - TS = -kT \ln Q \tag{4.37}$$

This is the so-called bridge equation for the canonical distribution, relating the macroscopic Helmholtz free energy F to the microscopic partition function Q. Since all the other thermodynamic quantities can readily be related to F, the statistical mechanical problem is now solved. For example

$$p = -\left(\frac{\partial F}{\partial V}\right)_T = kT \left(\frac{\partial \ln Q}{\partial V}\right)_T$$

$$= \frac{-\sum_n \sum_{\{j\}} \exp(-\beta E_n) \frac{dE_n}{dV}}{Q} \tag{4.38}$$

which is a special case, for the canonical distribution, of equation (2.12). Also, from equation (4.28),

$$U = \sum_n \sum_{\{j\}} P_{n\{j\}} E_n$$

$$= \frac{\sum_n \sum_{\{j\}} E_n \exp\left(-\frac{E_n}{kT}\right)}{\sum_n \sum_{\{j\}} \exp\left(-\frac{E_n}{kT}\right)} \quad (4.39)$$

The sum in the denominator is Q, and, by differentiation with respect to temperature, the sum

$$\sum_n \sum_{\{j\}} (E_n/kT^2) \exp(-E_n/kT)$$

is generated which, apart from the factor kT^2, is equal to the numerator. Thus

$$U = kT^2 \frac{1}{Q} \left(\frac{\partial Q}{\partial T}\right)_V = kT^2 \left(\frac{\partial \ln Q}{\partial T}\right)_V \quad (4.40)$$

Before we go on to discuss these results and the probability distribution, we will digress slightly to discuss the practicality of our approach. We assumed that we could measure the energy U of our system, and equation (4.28) is used to determine the Lagrange multiplier β in terms of U. In practice, the measurement of U would be very difficult, and what is usually measured is the temperature. We have identified the parameters β with $1/kT$, hence an alternative procedure would be to specify the temperature or alternatively β and use equation (4.28) or (4.40) to determine U. Thus, although in deriving our equations we require to know U, having derived them we can carry out actual calculations by knowing the temperature rather than the energy.

To return now to the probability function that we have derived. It is not obvious that the probability for the system having a particular energy has the peaked structure that we require. It is therefore necessary to calculate the possible deviations in energy, and this is conveniently done by finding the mean square deviation, that is,

$$\langle \Delta E \rangle^2_{\text{rms}} \equiv \langle (E - \langle E \rangle)^2 \rangle = \langle E^2 \rangle - 2E\langle E \rangle + \langle E \rangle^2 \rangle = \langle E^2 \rangle - \langle E \rangle^2$$

$$= \frac{1}{Q} \sum_n \sum_{\{j\}} E_n^2 \exp\left(-\frac{E_n}{kT}\right) - \left[\frac{1}{Q} \sum_n \sum_{\{j\}} E_n \exp\left(-\frac{E_n}{kT}\right)\right]^2$$

$$= kT^2 \left(\frac{\partial U}{\partial T}\right)_V = kT^2 C_V \quad (4.41)$$

where C_V is the heat capacity of the system at constant volume and we use the identification $U \equiv \langle E \rangle$. The final step can easily be checked by direct differentiation with respect to T of

$$U = \langle E \rangle = \frac{1}{Q} \sum_n \sum_{\{j\}} E_n \exp\left(-\frac{E_n}{kT}\right)$$

Thus we obtain for the fractional deviation

$$\frac{\langle \Delta E \rangle_{\text{rms}}}{U} = \frac{kT}{U} \left(\frac{C_V}{k}\right)^{\frac{1}{2}} \qquad (4.42)$$

Now it is usually the case that C_V and U are proportional to the number N of particles, so that the fraction deviation vanishes as N tends to infinity. An exception to this occurs near the critical point where C_V can be very greatly in excess of Nk. We would expect therefore the fluctuations to become large at this point, since all values of the energy become equally probable. Such a situation is found to occur experimentally.

Thus we have established the fundamental assumption on which all of our results so far have been based, namely that, at least for energy, the probability function is a sharply peaked function and that all values of the energy within that peak are macroscopically indistinguishable, i.e., from equation (4.42) they can deviate by amount of order $1/N^{\frac{1}{2}}$, where $N \sim 10^{24}$.

We have arrived at the result in this section, identical to the result for the isolated system, that, apart from a scaling constant, the uncertainty is to be identified with the entropy. Thus our discussion of entropy in Section 4.2 applies equally to the canonical distribution. The identification of the constant k with Boltzmann's constant can be carried out in analogous manner; the actual calculation will be performed in Chapter 9 and we shall assume that this has been done. We can obtain one further useful result. To obtain the canonical distribution we calculated the maximum value of H subject to the constraints

$$\sum_n \sum_{\{j\}} P_{n\{j\}} = 1 \quad \text{and} \quad \sum_n \sum_{\{j\}} P_{n\{j\}} E_n = U$$

This means that we maximize the function

$$K = H - \beta \sum_n \sum_{\{j\}} P_{n\{j\}} E_n - \lambda \sum_n \sum_{\{j\}} P_{n\{j\}} \qquad (4.43)$$

with no constraints.

However, with the benefit of hindsight we can interpret the maximization of equation (4.43) in a different way. If we put $\beta = 1/kT$ in this equation we can interpret the maximization procedure as maximizing the function

$$\left\{H - \frac{1}{kT}\sum_n \sum_{\{j\}} P_{n\{j\}} E_n\right\} \tag{4.44}$$

subject to the constraint

$$\sum_n \sum_{\{j\}} P_{n\{j\}} = 1$$

Since we have chosen $\beta = 1/kT$, the result of this procedure will be exactly the same as that given by our previous procedure, i.e., $P_{n\{j\}}$ will be given by equation (4.29). The corresponding maximum value of equation (4.44) will be

$$\left\{H - \frac{1}{kT}\sum_n \sum_{\{j\}} P_{n\{j\}} E_n\right\}_{max} = \left\{\frac{S}{k} - \frac{U}{kT}\right\} \tag{4.45}$$

where we have used equations (4.28) and (4.34). Hence using equation (4.37)

$$\left\{H - \frac{1}{kT}\sum_n \sum_{\{j\}} P_{n\{j\}} E_n\right\}_{max} = -\frac{F}{kT} \tag{4.46}$$

Now since the quantity in curly brackets is a maximum at a given volume and temperature, it follows that, because of the minus sign, F is a minimum at a given temperature and volume.

As an example of the use of this result consider a gas in which the molecules can undergo the reaction:

$$A + B \rightleftharpoons AB \tag{4.47}$$

Let the number of A, B and AB molecules be N_A, N_B and N_{AB} respectively subject to the constraint

$$N_A + N_B + 2N_{AB} = N$$

i.e., the total number of A and B molecules, whether dissociated or not, is a constant. For simplicity we shall assume that $N_A = N_B$ and therefore

$$2(N_A + N_{AB}) = N \tag{4.48}$$

Now the energy levels of the system will be functions of N_A only, since $N_{AB} = N/2 - N_A$, and similarly the probability of a state will also depend on N_A. Consequently the given mean energy will depend on the value of N_A. The procedure for such a problem would be to construct a free energy $F(N_A)$ for a given N_A using the canonical distribution function. To find the equilibrium value of N_A we would, according to the result discussed above, minimise the $F(N_A)$ with respect to N_A and hence find the equilibrium value of N_A. Substituting this value of N_A into $F(N_A)$ would give us the equilibrium free energy and hence all the other thermodynamic properties. This situation is discussed in further detail in Section 9.4.

4.4 System in a heat-and-particle bath (grand canonical distribution)

The final type of boundary conditions we shall consider for our system is one in which the system cannot only exchange *energy* with the large heat bath but also *particles*. As for the case of energy, the particle bath is assumed to be so large that it can supply or absorb particles from the system without changing its own thermodynamic state. Let us consider the case first where the system has a well-defined number of particles N. Then from our discussion in the previous section it is clear that the system will have accessible to it all states with N particles in them. If we now allow the number of particles to vary, then the states accessible to the system increase to include states with any number of particles in it. To define this thermodynamic state we have to specify three macroscopic parameters, instead of two. (Compare the discussion in Chapter 2 on systems with variable numbers of particles.) We shall choose these to be the volume V, the observed energy U (to be equated to the mean energy $\langle E \rangle$) and the observed number of particles $\mathcal{N}$ (equated to the mean number $\langle N \rangle$).

We introduce the probability $P_{Nn\{j\}}$ of the system having N particles and being in the quantum state $(n, \{j\})$ and determine its most likely value by maximizing the uncertainty

$$H = - \sum_{N=0}^{\infty} \sum_n \sum_{\{j\}} P_{Nn\{j\}} \ln P_{Nn\{j\}} \qquad (4.49)$$

subject to the constraints

$$\sum_{N=0}^{\infty} \sum_n \sum_{\{j\}} P_{Nn\{j\}} = 1 \qquad (4.50)$$

$$\sum_{N=0}^{\infty} \sum_n \sum_{\{j\}} P_{Nn\{j\}} E_n(N) = U \qquad (4.51)$$

and

$$\sum_{n=0}^{\infty} \sum_{n} \sum_{\{j\}} P_{Nn\{j\}} N = \mathcal{N} \qquad (4.52)$$

This problem is a natural extension from the cases discussed in Section 3.2. Indeed it is just like Question 3.7 except that an extra Lagrange multiplier, γ, is required. The result for the probability is

$$P_{Nn\{j\}} = \frac{\exp[-\gamma N - \beta E_n(N)]}{\Xi} \qquad (4.53)$$

and for the maximum uncertainty

$$H_m = \ln \Xi + \beta U + \gamma \mathcal{N} \qquad (4.54)$$

where

$$\Xi = \sum_{N=0}^{\infty} \sum_{n} \sum_{\{j\}} \exp[-\gamma N - \beta E_n(N)]$$

$$= \sum_{N=0}^{\infty} \sum_{n} g_n(N) \exp[-\gamma N - \beta E_n(N)]$$

$$= \sum_{N=0}^{\infty} e^{-\gamma N} Q(N) \qquad (4.55)$$

and $g_n(N)$ is the degeneracy of the energy level E_n of the system with N particles. β and γ are the Lagrange undetermined multipliers to be determined from equations (4.51) and (4.52). The probability distribution is known as the Gibbs grand canonical distribution and Ξ (pronounced as 'ksi') as the grand canonical partition function.

We follow the now familiar pattern of keeping the volume and the observed number of particles N in our system constant and altering the parameters of our heat bath such that a small amount of heat flows reversibly into the system. The corresponding change in the uncertainty is

$$\delta H_m = -\sum_{N=0}^{\infty} \sum_{n} \sum_{\{j\}} \delta(P_{Nn\{j\}} \ln P_{Nn\{j\}})$$

$$= +\beta \sum_{N=0}^{\infty} \sum_{n} \sum_{\{j\}} E_n(N) \delta P_{Nn\{j\}} \qquad (4.56)$$

where we have used the fact that

$$\sum_{N=0}^{\infty} \sum_{n} \sum_{\{j\}} \delta P_{Nn\{j\}} = 0 \qquad (4.57)$$

SYSTEM IN A HEAT-AND-PARTICLE BATH   49

and also, since the observed number of particles stays constant,

$$\sum_{N=0}^{\infty} \sum_{n} \sum_{\{j\}} N \delta P_{Nn\{j\}} = 0 \tag{4.58}$$

Now the corresponding change in the energy under these conditions is from (4.51)

$$\delta U = \sum_{N=0}^{\infty} \sum_{n} \sum_{\{j\}} E_n(N) \delta P_{Nn\{j\}} \tag{4.59}$$

Hence using equations (4.56) and (4.59), and the thermodynamic identity at constant volume and constant number of particles $T\delta S = \delta U$, we get

$$\delta S = \frac{\delta H_m}{\beta T} \tag{4.60}$$

which is the same result as for the canonical distribution, and with similar arguments to those which led to (4.34) and (4.35) we obtain

$$S = k H_m \tag{4.61}$$

and

$$\beta = \frac{1}{kT} \tag{4.62}$$

We still have to identify the Lagrange multiplier γ. Let us consider a small change δN in the observed number of particles in the system, this change taking place in such a way that the internal energy and the volume of the system remain constant. The corresponding change in the maximum uncertainty is

$$\delta H_m = -\sum_{N=0}^{\infty} \sum_{n} \sum_{\{j\}} \delta(P_{Nn\{j\}} \ln P_{Nn\{j\}})$$

$$= \sum_{N=0}^{\infty} \sum_{n} \sum_{\{j\}} \delta P_{Nn\{j\}} [\gamma N + \beta E_n(N)] \tag{4.63}$$

where we have used equation (4.53) for $P_{Nn\{j\}}$. The last term in this expression is β times the change in internal energy, which by definition is zero, and the first term is just the change in the number of particles. Since $S = kH_m$ we have

$$\delta S = k\gamma \delta N. \tag{4.64}$$

Comparison with the thermodynamic identity at constant volume and constant internal energy, $T\delta S = -\mu \delta \mathcal{N}$ gives

$$\gamma = -\frac{\mu}{kT} = -\beta\mu \qquad (4.65)$$

where μ is the chemical potential.

Our final expression for the probabilities is obtained from equation (4.53) using equations (4.62) and (4.65)

$$P_{Nn\{j\}} = \frac{\exp\left(\frac{[\mu N - E_n(N)]}{kT}\right)}{\Xi} \qquad (4.66)$$

The entropy follows from equations (4.61), (4.49) and (4.66)

$$S = k\ln\Xi + \frac{U}{T} - \frac{\mu\mathcal{N}}{T} \qquad (4.67)$$

If we define a thermodynamic potential

$$Z = \mu\mathcal{N} + TS - U \qquad (4.68)$$

then

$$Z = kT\ln\Xi \qquad (4.69)$$

If we can equate $\mu\mathcal{N}$ to the Gibbs free energy, then $Z = pV$, since by definition $G = U - TS + pV$. In the majority of cases this is a legitimate procedure and merely expresses the fact that G is an extensive quantity; we shall however meet two cases, one in Chapter 8 (the paramagnetic crystal) and one in Chapter 11 (the Bose gas), where this is not legitimate. If we follow standard thermodynamic practice and differentiate (4.68) and use the relation

$$TdS + \mathcal{N}d\mu = dU + pdV$$

and find that

$$dZ = \mathcal{N}d\mu + SdT + pdV$$

from which we get the thermodynamic relations

$$\mathcal{N} = \left(\frac{\partial Z}{\partial \mu}\right)_{T,V} \qquad (4.70)$$

$$S = \left(\frac{\partial Z}{\partial T}\right)_{\mu, V} \tag{4.71}$$

and
$$p = \left(\frac{\partial Z}{\partial V}\right)_{\mu, T} \tag{4.72}$$

Finally from the definition (4.68) of Z, we find for the internal energy

$$U = \mu \mathcal{N} + TS - Z \tag{4.73}$$

Thus the bridge equation (4.68) and the thermodynamic relations (4.70) to (4.73) enable us to calculate all the equilibrium macroscopic properties of the system. Sometimes it may be easier to calculate the pressure not from (4.72) but from the relation derived in Chapter 2, i.e.,

$$pV = \alpha U \tag{4.74}$$

when the volume dependence of the system's energy levels is of the form $V^{-\alpha}$.

Where the relation $G = \mu \mathcal{N}$ is valid then we run into an apparent contradiction since the bridge equation (4.69) takes the form

$$pV = kT \ln \Xi \tag{4.75}$$

which is not the same as (4.72). However comparison of these two expressions for p shows that $\ln \Xi$ must be of the form:

$$\ln \Xi = \varphi(\mu, T) V \tag{4.76}$$

where φ is a function of μ, T only, which ensures that all the extensive and intensive thermodynamic quantities have the correct dependence on N and V, for example,

$$\begin{aligned}S &= k \left[\frac{\partial [\varphi(\mu, T) T]}{\partial T}\right]_{\mu} V \\ &= k \left[\frac{\partial [\varphi(\mu, T) T]}{\partial T}\right]_{\mu} \mathcal{N}/\rho \end{aligned} \tag{4.77}$$

where ρ is the number density. Clearly S is extensive but on the other hand

$$p = kT\varphi(\mu, T) \tag{4.78}$$

is clearly intensive. What equation (4.76) is really saying is that in any calculation we must find the given volume dependence, if we are to obtain the correct N dependence for the macroscopic thermodynamic parameters. In fact it is not difficult to show that equation (4.76) has been forced upon us by the assumption that the Gibbs free energy is extensive, i.e. $G = \mu \mathcal{N}$ where μ is independent of $\mathcal{N}$.

It should also be noted that, as for the canonical distribution, equation (4.73) which determines T for given U may alternatively be used to determine U for given T, this latter interpretation being the one which is used in practice. Similarly equation (4.70) which is an equation for determining μ for a given $\mathcal{N}$, may alternatively be interpreted as an equation for determining $\mathcal{N}$ for a given μ. In practice, however, for this equation the first interpretation is usually used.

Since N can now take all values from 0 to ∞ we must consider the sharpness of the distribution of particle number around $\mathcal{N}$. The approach is very like that adopted for energy deviations in the canonical distribution and it is shown in Appendix 2 that

$$\frac{\langle \Delta N \rangle^2_{\text{rms}}}{\mathcal{N}^2} = \frac{kT}{\mathcal{N}^2} \left(\frac{\partial \mathcal{N}}{\partial \mu} \right)_{V,T} = \frac{kT}{B\mathcal{N}} \left(\frac{\mathcal{N}}{V} \right) \tag{4.79}$$

where B is the isothermal bulk modulus. This result shows immediately that provided B is finite, the fractional deviation vanishes like $1/\mathcal{N}^{1/2}$. An exceptional situation is found near critical points where B tends to zero and the fluctuations are significant.

At this stage one may digress to consider what physical system we have in mind in which the number of particles can change. Apart from systems involving chemical reactions, there is another which fits the condition. Consider a macroscopic system in a heat bath and consider also a small subsystem of that system. The situation is illustrated schematically in Figure 4.1. The sub-system has precisely the boundary condition that we have imposed on our system in the grand canonical distribution, the rest of the system acting as a heat and particle bath. Provided the volume of the sub-system is large on a microscopic scale (but small compared with the system) the energy of interaction between the sub-system and the rest of the system will be small compared with the energy of the sub-system. Thus we can apply the results of the grand canonical distribution to the sub-system and calculate its thermodynamic properties.

On the other hand we could consider the whole system and apply the results of the canonical ensemble to it. In both cases we could expect to get the same results for all the intensive parameters and the extensive parameters to be in the ratio of the number of particles in the sub-system to the number in the whole system. That this must be so follows from our considerations that for the grand canonical distribution the probability of the system having a number of particles in it macroscopically distinguishable from $\mathcal{N}$ is vanishingly small. Thus, to all intents and purposes, the system has a fixed number of particles. It follows therefore that we can use either the canonical or grand canonical ensemble to calculate the thermodynamic properties.

The exception to this rule would be if the system were in the region of a critical point. What does this mean in terms of the sub-system we have been discussing? Now the advantage of considering the sub-system is that we can easily make macroscopic copies of it. We simply consider another part of the system having the same volume as our original sub-system, this then is our macroscopic copy. Clearly we can repeat this several times over, and the situation is illustrated in Figure 4.1. If we are near a critical point, then the probability of observing macroscopically different values of the number of particles in each sub-system becomes large. For the whole system this means that the density is different at different points in the system, i.e., we have density fluctuations on a macroscopic scale. Such fluctuations are observed experimentally and are responsible for the condition known as critical opalescence.

Finally in this section we shall obtain a variational statement of thermal equilibrium as we did for the canonical ensemble. The function which we had to minimize to obtain the grand canonical distribution was

$$K = -\sum_{N=0}^{\infty} \sum_{n} \sum_{\{j\}} \{P_{Nn\{j\}} \ln P_{Nn\{j\}} + \beta P_{Nn\{j\}} E_n(N) + \gamma P_{Nn\{j\}} N + \lambda P_{Nn\{j\}}\} \tag{4.80}$$

By similar reasoning as for the canonical ensemble we can interpret this as maximizing the function

$$\left\{ H - \frac{1}{kT} \sum_{N=0}^{\infty} \sum_{n} \sum_{\{j\}} P_{Nn\{j\}} (E_n - \mu N) \right\} \tag{4.81}$$

subject to the constraint

$$\sum_{N=0}^{\infty} \sum_{n} \sum_{\{j\}} P_{Nn\{j\}} = 1$$

where we have put $\beta = 1/kT$ and $\gamma = -\mu/kT$. Using equations (4.61), (4.51) and (4.52) we find the maximum value of this function to be

$$\left\{ H - \frac{1}{kT} \sum_{N=0}^{\infty} \sum_{n} \sum_{\{j\}} P_{Nn\{j\}} (E_n - \mu N) \right\}_{\max}$$

$$= \frac{S}{k} - \frac{U}{kT} + \frac{\mu \mathcal{N}}{kT}$$

$$= \frac{pV}{kT} \tag{4.82}$$

where we have used (4.68) with $Z = pV$. Thus at constant V, T and μ, p is a maximum at equilibrium. This result is not as easily applicable as the corresponding result for the canonical distribution. This is because μ rather than N is to be held constant and we do not propose to discuss any examples of this.

CHAPTER 4: SUMMARY OF MAIN POINTS

1. The application of the law of maximum uncertainty applied to a thermally isolated system, leads to the microcanonical distribution.
2. The relation between the macroscopic quantities and the microscopic ones, is simply expressed by the formula $S = k \ln \Omega$.
3. The microcanonical partition function Ω is the number of quantum mechanical states available to the isolated system.
4. For a system in a heat bath, the appropriate distribution is the canonical distribution.
5. The canonical partition function is

$$Q = \sum_n \sum_{\{j\}} \exp(-\beta E_n)$$

6. For a system in a heat and particle bath, the appropriate distribution is the grand canonical distribution.
7. For that case $Z = \mu \mathcal{N} + TS - U = kT \ln \Xi$. If the Gibbs free energy $G = \mu \mathcal{N}$ then, $pV = kT \ln \Xi$.
8. The grand canonical partition function is

$$\Xi = \sum_N \exp(\mu N) Q(N)$$

9. For almost all cases, the fluctuations in the values of the macroscopic variables are found to be proportional to the reciprocal of the square root of the number of particles.
10. The exception to this is near a critical point, where macroscopic fluctuations can occur. These fluctuations are observed experimentally.

QUESTIONS

4.1 Consider the situation where we have two systems and let the probability of system A being in a state $(n, \{j\})$ be $P^{(A)}_{n\{j\}}$ and the system B being in a state $(m, \{k\})$ be $P^{(B)}_{m\{k\}}$. Consider first the situation where the two systems are weakly interacting so that the joint probability $P_{n\{j\},m\{k\}}$ of system A being in state $(n, \{j\})$ and system B in state $(m, \{k\})$ is given by

$$P_{n\{j\},m\{k\}} = P^{(A)}_{n\{j\}} P^{(B)}_{m\{k\}}$$

Hence show that the entropy of the combined systems S is equal to the sum of the entropies S_A, S_B of systems A and B.

If the systems are not interacting significantly show that

$$S - (S_A + S_B) = k \sum_{n\{j\}} \sum_{m\{k\}} P_{n\{j\},m\{k\}} \ln \left[\frac{P^{(A)}_{n\{j\}} P^{(B)}_{m\{k\}}}{P_{n\{j\},m\{k\}}} \right]$$

By making use of the numerical inequality $\ln x \leqslant (x - 1)$ show that $S \leqslant S_A + S_B$.

4.2 Look up the derivation of the canonical distribution and use similar methods to deal with another situation in which the walls dividing the systems from the reservoir are permeable to energy and to volume (i.e., the walls are flexible). If the mean energy and volume are U and V respectively, show that the most probable distribution for the system in a state $(n, \{j\})$ and with V_α is

$$P_{n\{j\},V_\alpha} = \frac{\exp(-\beta E_n - \gamma V_\alpha)}{Z}$$

where

$$Z = \sum_n \sum_\alpha \exp(-\beta E_n - \gamma V_\alpha)$$

and β and γ are given by

$$U = -\left(\frac{\partial \ln Z}{\partial \beta} \right)_\gamma \quad \text{and} \quad V = -\left(\frac{\partial \ln Z}{\partial \gamma} \right)_\beta$$

By differentiating the maximum uncertainty regarded as a function of U and V and comparing it with the thermodynamic equation

$$T dS = dU + p dV$$

show that $\beta = 1/kT$ and $\gamma = \beta p$. Hence show that $G = -kT \ln Z$ and furthermore that for a given pressure and temperature G is a minimum.

4.3 A physical system has three types of particles, A, B and C, which undergo the chemical reaction $A \rightleftharpoons B + C$. If the mean number of particles N_A, N_B and N_C and the total mean energy are given, find the most probable distribution.

By comparing with the thermodynamic equation

$$\mu_A dN_A + \mu_B dN_B + \mu_C dN_C + TdS = dU + pdV$$

establish a connection between the quantities appearing in this equation and the normalizing factor of the most probable distribution function.

4.4 Find an expression for the fractional deviation in *energy* in the grand canonical distribution.

4.5 For the canonical distribution the probability of a system having a particular energy is given by

$$P(E_n) = \frac{g_n \exp(-\beta E_n)}{Q}$$

Assuming that g_n varies as E_n^α where α is between 0 and 1, estimate for, say, two values of α the ratio of $P(E_n = U)$ and $P(E_n = 100\,U)$. If you were to interpret probability as the ratio of the number of times an event occurred in a very large number of identical systems to the number of such systems, how many of the systems would have an energy $100\,U$ if one thousand of the systems had an energy U? How many identical systems would you need to have if you were to have a reasonable chance of finding one system with an energy $100\,U$?

4.6 In Chapters 10 and 13 we shall be discussing real gases, developing methods which show that for some gases a reasonable approximation to the canonical function is

$$Q = \text{const} \left(\frac{V - Nb}{N}\right)^N \left(\frac{2\pi mkT}{h^2}\right)^{3N/2} \exp\left(\frac{N^2 a}{VkT}\right)$$

where a and b are small positive constants. Show that this leads to Van der Waals equation of state for a gas, and calculate the mean energy.

4.7 A system consists of five labelled particles (A, B, C, D and E), each of which can be in non-degenerate single-particle states with energies $0, \epsilon, 2\epsilon$. If the system has an energy $(3\frac{1}{2} \pm \frac{1}{2})\epsilon$ (where the range is determined by instrumental limitations), what are the mean numbers of particles in each of the energy states?

4.8 Assuming that for a particular system F is a known function of V and T, show that

$$p = -\left(\frac{\partial F}{\partial V}\right)_T$$

and find expressions for S, C_V, C_p, U, G and H.

4.9 This question makes use of the microcanonical distribution to deal with an idealised model of rubber elasticity which can be explicitly solved. A one-dimensional chain consists of N links each of length a, and each link may lie in the forward or reverse alignment. Let the distance between the ends of the chain be x (note that the greatest possible value of x is Na, corresponding to a configuration in which all elements are aligned in the forward direction).

(a) Show that the numbers of links in the forward and reverse alignments are, respectively, $\frac{1}{2}(N + x/a)$ and $\frac{1}{2}(N - x/a)$.

(b) Prove that the number of ways of arranging the chain so as to give a distance x is

$$\frac{N!}{[\frac{1}{2}(N + x/a)]![\frac{1}{2}(N - x/a)]!}$$

(c) Use Stirling's approximation for the three large numbers N, $\frac{1}{2}(N + x/a)$ and $\frac{1}{2}(N - x/a)$ to obtain a simplified (i.e., without factorials) expression for the entropy S of the chain.

(d) By use of the thermodynamic result that the tension in the chain equals $-T(dS/dx)$ (since in our idealization $U = 0$), prove that the tension is

$$\frac{kT}{2a} \ln \left(\frac{1 + x/Na}{1 - x/Na} \right)$$

(e) Indicate without detailed calculation how the approach would have to be modified if the forward and reverse link alignments were associated with energies $-\epsilon$ and $+\epsilon$ respectively.

4.10

The figure above shows a section through a monatomic crystal of N atoms, the crosses representing the normal position of the atoms. At high temperatures some atoms may leave their normal positions, thus creating vacancies, and move to interstitial positions indicated by dots. If an atom is in an interstitial position it has an additional energy ϵ over its energy in its normal position. If we take the zero of energy of the configuration when all atoms are in their normal position, show that the energy for any other configuration is

$$E_n = n\epsilon$$

where n is the number of interstitials. Show that for a given n the number of possible configurations is

$$\left[\frac{N!}{n!(N-n)!}\right]^2$$

(remembering that there are N normal positions and N interstitial positions). These two results give us the 'energy levels' of the system and the degeneracies of each level; hence show that the probability of the system having an energy $n\epsilon$ is given by

$$P(n\epsilon) = \left[\frac{N!}{n!(N-n)!}\right]^2 \left(\frac{e^{-\beta n\epsilon}}{Q}\right)$$

where

$$Q = \sum_{N=0}^{N} \left[\frac{N!}{n!(N-n)!}\right]^2 e^{-\beta n\epsilon}$$

By making use of Stirling's approximation find the mean value of $n\epsilon$, and hence n, and show that the probability distribution is sharply peaked about this value.

4.11 The Einstein model of an N-atom solid considers $3N$ non-interacting one-dimensional simple haromonic oscillators, each of which may have an energy (relative to its zero-point energy) of $n\epsilon$ where n is zero or any positive integer. Prove that the *microcanonical* partition function is

$$\Omega = \frac{(3N - 1 + E/\epsilon)!}{(3N - 1)!(E/\epsilon)!}$$

where E is the total energy. Use the bridge equation and the general thermodynamic result that

$$\frac{1}{T} = \left(\frac{\partial S}{\partial E}\right)_N$$

to find an expression for the energy $E(T)$ and hence show that the heat capacity is given by

$$C = 3Nk \left(\frac{\epsilon}{kT}\right)^2 \frac{\exp(\epsilon/kT)}{[\exp(\epsilon/kT) - 1]^2}$$

CHAPTER 5
Partition Functions of Simple Sub-systems

5.1 Introduction
5.2 Particle with two possible states with energies ϵ_1 and ϵ_2
5.3 One-dimensional simple harmonic oscillator (SHO)
5.4 Rotating heteronuclear diatomic molecule
5.5 Particle-in-a-box
5.6 One heteronuclear diatomic molecule in a box
5.7 Two weakly interacting distinguishable molecules in a box
5.8 Conclusion
Questions

5.1 Introduction

We have shown in Chapter 4 that, with one exception, the results from statistical mechanics are independent of the boundary conditions of the system. This is not as surprising as it at first seems. Consider a cubic container of side L. The surface area is equal to $6L^2$, whereas the volume is equal to L^3, so we might expect the relative size of surface effects to volume effects to be approximately equal to $6/L = 6(n/N)^{1/3}$, where n is the particle number density. In the limit of large N, which is the thermodynamic limit, the surface effects are negligible. The exception, mentioned above, is in the neighbourhood of a critical point. We shall see in the next chapter, that the Einstein solid which we have already considered in Chapter 4 in the microcanonical distribution, has the same results when considered in the canonical distribution. Since this simple model does not possess a critical point, the results are valid over the whole temperature range. Similarly, in Chapter 7, we shall consider a simple model of a paramagnetic solid in both the canonical and the grand canonical distributions; the results in both cases are identical. Again this model does not possess a critical point. Other results of a similar nature will appear in Chapters 9 and 10 for gases, and for photons in Chapter 11. The choice of distribution is therefore a pragmatic one: whichever is mathematically easier we shall use (emphasizing again the proviso, that in the neighbourhood of critical points we must exercise caution). It turns out that for this reason we shall not use the microcanonical distribution, except for the example already given in Chapter 4. The reason for this is that the constraint that the total energy be given, albeit within limits, makes it mathematically more difficult than, say, the canonical distribution, where only the mean energy is given.

However, it must be emphasized that, if the canonical partition function is easier to calculate than the microcanonical one, it is only relatively so. For any real system the evaluation of the canonical partition function

$$Q = \sum_n \sum_{\{j\}} \exp(-\beta E_n) = \sum_n g_n \exp(-\beta E_n) \tag{5.1}$$

represents a formidable task. For a normal macroscopic system the number of particles N is of the order of 10^{27}, all of which are interacting with each other. We are always reduced to some approximate method for a real system. Replacing that system by a model system which is capable of solution either exactly or to a good approximation. In fact, much of the physics of the problem is trying to find a reasonable model of the real world situation. Some examples of these are given in later chapters. At this stage in the argument, we do not wish to confuse the calculation procedure with the physics of obtaining a soluble model; what we want to do now, is actually to calculate a canonical partition function. The simplest partition function to calculate is that for a single atom or a single molecule. This is because the energy levels and degeneracies of such sub-systems (we refer to them as sub-systems because they are constituents of the later more realistic systems we shall consider) can be calculated exactly. To differentiate between the realistic partition function for a macroscopic number of particles and these artificial functions, we shall denote them and their subsequent 'thermodynamic' properties by lower case letters, for example, the partition function will be written

$$q = \sum_n g_n \exp(-\beta \epsilon_n) \tag{5.2}$$

for a sub-system whose states $(n, \{j\})$ have energies ϵ_n with degeneracies g_n.

In choosing to discuss such systems because of the ease of their calculation, we will have to abandon the arguments about the sharpness of the distribution. Thus, for example, the energy of the system will have large fluctuations. We can, since we are considering the canonical distribution still have a meaning for temperature; it will be the temperature of the heat bath within which the system is sitting. Nevertheless the situation is artificial and if the only consideration were to illustrate the method of calculation, the game would not be worth a candle. However, it turns out that we will be able to use these results in the calculation of the partition function of more realistic model systems. We start, therefore, by considering the simplest possible system, that of a particle with only two accessible energy levels, each of which is non-degenerate.

5.2 Particle with two possible states with energies ϵ_1 and ϵ_2

This might describe two electronic energy levels in a nitric oxide (NO) molecule, separated by 2.45×10^{21} joules, or an atomic magnetic moment associated with spin angular momentum $\frac{1}{2}\hbar$, capable of two possible orientations with respect to an applied field B. In the latter case, $\epsilon_1 = -\gamma B$ and $\epsilon_2 = +\gamma B$ where γ denotes the magnetic moment.

For these cases,

$$g_1 = g_2 = 1 \tag{5.3}$$

and

$$\begin{aligned} q &= \sum_n \exp\left(-\frac{\epsilon_n}{kT}\right) \\ &= \exp\left(-\frac{\epsilon_1}{kT}\right) + \exp\left(-\frac{\epsilon_2}{kT}\right) \end{aligned} \tag{5.4}$$

Hence, using equation (4.37) with lower-case letters for sub-system functions,

$$\begin{aligned} f &= -kT\ln q \\ &= -kT\ln\left[\exp\left(-\frac{\epsilon_1}{kT}\right) + \exp\left(-\frac{\epsilon_2}{kT}\right)\right] \end{aligned} \tag{5.5}$$

$$= \epsilon_1 - kT\ln\left[1 + \exp\left(-\frac{\Delta\epsilon}{kT}\right)\right] \tag{5.6}$$

where $\Delta\epsilon = (\epsilon_2 - \epsilon_1) > 0$ since by definition $\epsilon_2 > \epsilon_1$. Hence, from thermodynamics,

$$s = -\left(\frac{\partial f}{\partial T}\right)_V$$

$$= k\ln\left[1 + \exp\left(-\frac{\Delta\epsilon}{kT}\right)\right] + k\frac{\frac{\Delta\epsilon}{kT}}{\exp\left(\frac{\Delta\epsilon}{kT}\right) + 1} \quad (5.7)$$

It is interesting to observe the behaviour of s in the high and low temperature limits:

At high T (i.e., $T \gg \Delta\epsilon/k$), $s = k\ln 2$ \quad (5.8)

This is just what we should expect, since at high temperatures the difference $\Delta\epsilon$ in energy between the levels is negligible, so that each level has probability $\frac{1}{2}$ of occupation which leads, in terms of equation (3.15), to an uncertainty of ln2 or entropy given by equation (5.8).

At low T (i.e., $T \ll \Delta\epsilon/k$), $s = 0$ \quad (5.9)

This is consistent with the third law. Also,

$$u = f - Ts$$

$$= \epsilon_1 + \frac{\Delta\epsilon}{\exp\left(\frac{\Delta\epsilon}{kT}\right) + 1} \quad (5.10)$$

and finally the heat capacity c is

$$c = \left(\frac{\partial u}{\partial T}\right)_V = k\left(\frac{\Delta\epsilon}{kT}\right)^2 \frac{\exp\left(\frac{\Delta\epsilon}{kT}\right)}{\left[\exp\left(\frac{\Delta\epsilon}{kT}\right) + 1\right]^2} \quad (5.11)$$

This function is zero in the low and high temperature limits and has a maximum value of $0.439k$ when $T = 0.417\,\Delta\epsilon/k$. We must remember that these results refer to a single molecule, but we shall find that *assemblies* of such molecules, for example, nitric oxide (Chapter 9) or a paramagnetic solid in a magnetic field (Chapter 7) do exhibit such heat capacity 'anomalies'. Of course for a single molecule we should not necessarily expect to find sharp distributions, a point which was regarded as essential in Chapter 1. To emphasize the slightly artificial device of dealing with a single molecule, let us calculate $\langle \Delta E \rangle_{\text{rms}}$ as a measure of fluctuations, and compare it with u to find the fractional deviation in energy. Using equation (4.42) we obtain

$$\frac{\langle\Delta E\rangle_{\text{rms}}}{u} = \frac{\Delta\epsilon}{\epsilon_1 \exp\left(\dfrac{\Delta\epsilon}{2kT}\right) + \epsilon_2 \exp\left(-\dfrac{\Delta\epsilon}{2kT}\right)} \quad (5.12)$$

Thus at high T (i.e. $T \gg \Delta\epsilon/k$) the fractional deviation is $\Delta\epsilon/(\epsilon_1 + \epsilon_2)$, which in general is not vanishingly small. However for $T \ll \Delta\epsilon/k$ it takes the form $(\Delta\epsilon/\epsilon_1)\exp(-\Delta\epsilon/2kT)$ which approaches zero as $T \to 0$. Now we have already noted that $s \to 0$ as $T \to 0$, so the uncertainty is also tending to zero. It is not difficult to see from this that, at the absolute zero, the sub-system will definitely be in the ground state. The uncertainty about its state is therefore zero. This in fact is a general result, for the canonical distribution and the other distributions, that at absolute zero the system is definitely in the ground state. However, the third law tells us that absolute zero is unattainable in a finite number of steps. Nevertheless, some systems have been taken very close – in the case of helium to about 10 microkelvin.

5.3 One-dimensional simple harmonic oscillator (SHO)

This might be expected to represent internal oscillations of molecules or oscillations of atoms in crystalline arrays. The energy levels of an SHO are obtainable from the Schrödinger equation and are given by

$$\epsilon_n = h\nu\left(n + \tfrac{1}{2}\right), \qquad n = 0, 1, 2, \ldots \quad (5.13)$$

where ν is the classical frequency. Hence

$$q = \sum_{n=0}^{\infty} \exp\left[-\frac{h\nu\left(n + \tfrac{1}{2}\right)}{kT}\right] \quad (5.14)$$

since $g_n = 1$. This is a geometric series which is explicitly summable:

$$\begin{aligned} q &= \frac{\exp\left(-\dfrac{h\nu}{2kT}\right)}{1 - \exp\left(-\dfrac{h\nu}{kT}\right)} \\ &= \tfrac{1}{2}\operatorname{cosech}\left(\frac{h\nu}{2kT}\right) \\ &= \tfrac{1}{2}\operatorname{cosech}\left(\frac{\theta_\nu}{2T}\right) \end{aligned} \quad (5.15)$$

where

$$\theta_\nu \equiv \frac{h\nu}{k} \tag{5.16}$$

Hence

$$f = -kT\ln q = kT\ln\left[2\sinh\left(\frac{\theta_\nu}{2T}\right)\right] \tag{5.17}$$

and

$$s = -\left(\frac{\partial f}{\partial T}\right)_V = k\left[\frac{\theta_\nu}{2T}\coth\left(\frac{\theta_\nu}{2T}\right) - \ln\left(2\sinh\frac{\theta_\nu}{2T}\right)\right] \tag{5.18}$$

At high T (i.e., $T \gg \tfrac{1}{2}\theta_\nu$), $s \to k\ln(T/\theta_\nu)$, and at low T (i.e., $T \ll \tfrac{1}{2}\theta_\nu$), $s = 0$, consistent with the third law. Also

$$u = f + Ts = \tfrac{1}{2}k\theta_\nu \coth\left(\frac{\theta_\nu}{2T}\right) \tag{5.19}$$

and

$$c = \left(\frac{\partial u}{\partial T}\right)_V = k\left(\frac{\theta_\nu}{2T}\right)^2 \operatorname{cosech}^2\left(\frac{\theta_\nu}{2T}\right) \tag{5.20}$$

In the high-temperature limit this expression reduces to $c = k$, and as $T \to 0$, $c \to 0$. Once more it is clear that fractional energy deviations are far from small. We find

$$\frac{\langle \Delta E \rangle_{\text{rms}}}{u} = \operatorname{sech}\left(\frac{\theta_\nu}{2T}\right) \tag{5.21}$$

Thus at high T the fractional deviation in energy is equal to unity, and at low T it tends to zero as $2\exp(-\theta_\nu/2T)$. However, it will appear later that when a system of many oscillators is considered, the magnitude of the fluctuations is reduced to a level where it may usually be entirely neglected at all temperatures.

5.4 Rotating heteronuclear diatomic molecule

If the molecule is regarded as a rigid rotator of moment of inertia I, then its quantum-mechanical energy levels are given by

$$\epsilon_J = \frac{\hbar^2}{2I} J(J+1) \tag{5.22}$$

with degeneracy

$$g_J = 2J + 1 \tag{5.23}$$

and

$$J = 0, 1, 2 \ldots$$

The new feature here is the degeneracy, meaning that there are $(2J + 1)$ states which have the same energy ϵ_J but are distinguishable in some other way (i.e., the z-component of their angular momentum) and therefore have different labels $\{j\}$. In evaluating the canonical partition we have to sum over states, bearing in mind that J only labels *energy* levels. Hence

$$q = \sum_{J=0}^{\infty} (2J + 1) \exp\left[-\frac{\hbar^2 J(J+1)}{2kTI}\right] \tag{5.24}$$

Unfortunately, this cannot be explicitly summed, so that in general one is faced with numerical computation. However, limiting cases can be extracted:

(a) At low $T (T \ll \theta_r; \theta_r \equiv \hbar^2/2Ik)$ the rotator is extremely likely to be found in the ground state $(J = 0)$, and this is reflected in the mathematics by observing that the series converges very rapidly. Taking only the first two terms

$$q \approx 1 + 3 \exp\left(-\frac{2\theta_r}{T}\right) \quad \text{when} \quad T \ll \theta_r$$

Hence, by similar arguments to those given above,

$$s = 3k \left(\frac{2\theta_r}{T}\right) \exp\left(-\frac{2\theta_r}{T}\right) \tag{5.25}$$

and

$$c = 3k \left(\frac{2\theta_r}{T}\right)^2 \exp\left(-\frac{2\theta_r}{T}\right) \tag{5.26}$$

Both of these expressions go to zero as $T \to 0$, in accordance with the third law.

(b) At high $T (T \gg \theta_r)$, the physical situation is that a wide range of states have comparable probabilities, and mathematically many small but almost equal terms contribute to the sum. In this case (and in similar cases which will frequently arise) it is possible to approximate the sum for q by an integral.

Strictly of course J only takes discrete values, but at high temperatures vast numbers of states have comparable (if small) probabilities of occupation so that a large range of J-values may be expected to make comparably significant contributions to the sum (5.24). In such a situation, the procedure is to treat J as a quasi-continuous variable, in the sense that ranges dJ may be very small compared with all J values except the few lowest, but at the same time include many J-values. For instance, suppose a million (10^6) terms are significant in the sum (5.24); then a choice of $dJ = 100$ includes 100 almost equal values and the transition to an integral is desirable and easy. (Note that in many-particle systems this approach can sometimes be invalidated if the ground-state alone is dominatingly occupied. This, however, is not the case here as can readily be established by comparing the first few terms of the sum in equation (5.24).)

So the procedure is to treat J as a quasi-continuous variable and to ask how many states there are with J-values in the range dJ. Since the degeneracy is $g_J = (2J + 1)$, the answer is $g_J dJ = (2J + 1)dJ$.

$$q \approx \int_0^\infty (2J + 1) \exp\left[-\frac{J(J + 1)\theta_r}{T}\right] dJ \quad \text{when} \quad T \gg \theta_r$$

This integral is easily reduced to a standard form

$$\int_0^\infty e^{-x} dx$$

and hence

$$q \approx \frac{T}{\theta_r} \quad \text{when} \quad T \gg \theta_r \tag{5.27}$$

Hence

$$s \approx k\left[1 + \ln\frac{T}{\theta_r}\right] \tag{5.28}$$

and

$$c \approx k \tag{5.29}$$

With modern microcomputers it is very easy to evaluate numerically the summations involved in calculating the thermodynamic quantities. Question 5.6 takes the reader through the numerical computation of the heat capacity. The result shows that at a temperature just below θ_r it has a slight maximum.

5.5 Particle-in-a-box

We consider a single particle in a cubical box of side $V^{1/3}$ where V is the volume. In fact the results are independent of the shape of the container, but our particular choice simplifies the notation. If the particle has only translational degrees of freedom, then the quantum-mechanical states of the sub-system are specified by three integer quantum numbers n_x, n_y, n_z, which in turn specify the three components of linear momentum, i.e.,

$$p_x = \frac{hn_x}{2V^{1/3}} \quad \text{with} \quad n_x = 1, 2, 3, \ldots$$

$$p_y = \frac{hn_y}{2V^{1/3}} \quad \text{with} \quad n_y = 1, 2, 3, \ldots \quad (5.30)$$

$$p_z = \frac{hn_z}{2V^{1/3}} \quad \text{with} \quad n_z = 1, 2, 3, \ldots$$

The possible energy levels of the system are therefore

$$\begin{aligned}\epsilon &= (p_x^2 + p_y^2 + p_z^2)/2m \\ &= k\theta_t(n_x^2 + n_y^2 + n_z^2)\end{aligned} \quad (5.31)$$

where $\theta_t \equiv h^2/8mkV^{2/3}$ is a characteristic temperature for translational motion. For a general ϵ there can be a high degree of degeneracy for such a sub-system, since there can be many different sets of three integers whose summed squares will produce the same result. Because of this and the problem associated with calculating the degeneracy, we will sum over all states irrespective of their energies (the alternative would be to sum over all states with the same energy, to get the degeneracy, and then over all energies). Thus,

$$q = \sum_{n_x}\sum_{n_y}\sum_{n_z} \exp[-(n_x^2 + n_y^2 + n_z^2)\theta_t/T] \quad (5.32)$$

Two observations prove of value here. Firstly, a general point which has wide consequences in later applications: since, given the form of the function to be summed, the sum may be done with any chosen order to terms, it would appear to be sensible to decide to fix n_x and n_y and sum

68 PARTITION FUNCTIONS OF SIMPLE SUB-SYSTEMS

first over all possible n_z. Having done that, n_x may remain fixed and the summation over all n_y performed. Finally, the remaining sum is over all n_x. Algebraically this means that

$$q = \left\{\sum_{n_x} \left[\exp\left(-n_x^2 \frac{\theta_t}{T}\right)\right]\right\} \times \left\{\sum_{n_y} \left[\exp\left(-n_y^2 \frac{\theta_t}{T}\right)\right]\right\}$$

$$\times \left\{\sum_{n_z} \left[\exp\left(-n_z^2 \frac{\theta_t}{T}\right)\right]\right\} \qquad (5.33)$$

$$= \left\{\sum_{n=1}^{\infty} \left[\exp\left(-n^2 \frac{\theta_t}{T}\right)\right]\right\}^3 \qquad (5.34)$$

Unfortunately, the sum still cannot be explicitly evaluated. However, and this is the second observation, for particles in macroscopic volumes, θ_t is very small indeed (for example, for a helium atom in a one-litre volume, $\theta_t = 6 \times 10^{-17} K$!), so that for *any* physically realizable low temperature we are actually dealing with a *high-temperature* limiting condition. Consequently, the sum is composed of a vast number of tiny terms so that, as with the high-temperature rotating molecule of Section 5.4, it may be converted to an integral and negligible error. As before, we treat n as a quasi-continuous variable and ask how many states there are with n-values in the range dn. The answer here is simply dn, so that when $T \gg \theta_t$

$$\sum_{n=1}^{\infty} \exp\left(-n^2 \frac{\theta_t}{T}\right) \approx \int_0^{\infty} \exp\left(-n^2 \frac{\theta_t}{T}\right) dn \qquad (5.35)$$

$$= \left(\frac{\pi T}{4\theta_t}\right)^{1/2} \qquad (5.36)$$

Thus

$$q = \left(\frac{\pi T}{4\theta_t}\right)^{3/2} = \left(\frac{2\pi m k T}{h^2}\right)^{3/2} V \qquad (5.37)$$

Hence

$$f = -(3/2)kT \ln\left(\frac{2\pi m k T}{h^2}\right) - kT \ln V \qquad (5.38)$$

Thus

$$p = -\left(\frac{\partial f}{\partial V}\right)_T = \frac{kT}{V} \qquad (5.39)$$

Also

$$s = -\left(\frac{\partial f}{\partial T}\right)_V = k\ln V + \tfrac{3}{2}k\ln T + \tfrac{3}{2}k\ln\left(\frac{2\pi mk}{h^2}\right) + \tfrac{3}{2}k \quad (5.40)$$

Any alarm at noticing that $s \to \infty$ as $T \to 0$ is dispelled by remembering that this is an exclusively high-temperature expression which therefore cannot be legitimately extrapolated to $T = 0$. The appropriate low temperature ($T \ll \theta_t$) partition function would be $q = \exp(-\theta_t/T)$, and this predicts an entropy satisfying the third law, but is otherwise rather unphysical because θ_t is usually an unattainably low temperature. Finally

$$c_V = T\left(\frac{\partial s}{\partial T}\right)_V = \tfrac{3}{2}k \quad (5.41)$$

It must be stressed that although this result and the equation of state $pV = kT$ look familiar, we are not dealing here with a perfect gas but only a single atom. Let us once again calculate the fractional energy fluctuation:

$$\frac{\langle\Delta E\rangle_{\text{rms}}}{u} = \frac{kT}{u}\left(\frac{c_V}{k}\right)^{1/2} = \left(\frac{2}{3}\right)^{1/2} = 0.816 \quad (5.42)$$

Thus, fluctuations are very nearly as large as the internal energy u itself at all temperatures $T \gg \theta_t$. However, it will be shown later than q is a most important factor in the form of Q for a macroscopic number of gas atoms.

5.6 One heteronuclear diatomic molecule in a box

The problem we raise here is how to write down a partition function for *one sub-system* capable of *several modes* of motion. For instance a diatomic molecule can translate, rotate, vibrate and may also allow electronic transitions. We know now how to deal with each case separately as in Sections 5.5, 5.4, 5.3 and 5.2 respectively. A solution of the appropriate Schrödinger equation would of course yield states characterized by several quantum numbers, and it is not obvious that these states would fall into independent categories of translation, rotation, vibration and electronic transition. To a first approximation, however, that is the case, i.e.,

$$\epsilon = \epsilon_t + \epsilon_r + \epsilon_v + \epsilon_e \quad (5.43)$$

To the extent that this is true, it is most convenient because then

70 PARTITION FUNCTIONS OF SIMPLE SUB-SYSTEMS

$$q = \sum \exp\left[-\frac{\epsilon_t + \epsilon_r + \epsilon_v + \epsilon_e}{kT}\right]$$

$$= \left[\sum \exp\left(-\frac{\epsilon_t}{kT}\right)\right]\left[\sum \exp\left(-\frac{\epsilon_r}{kT}\right)\right]$$

$$\times \left[\sum \exp\left(-\frac{\epsilon_v}{kT}\right)\right]\left[\sum \exp\left(-\frac{\epsilon_e}{kT}\right)\right] \quad (5.44)$$

This last step is allowable for the same reasons as the similar case in Section 5.5; namely, that we are free to choose to perform the total summations in steps, dealing first with translational states, then rotational and so on. (Note that it would not be possible if the modes *interacted* in the sense that, say, vibrational states were influenced by a particular rotational state which the molecule happened to be in.) Thus

$$q = q_t q_r q_v q_e \quad (5.45)$$

where the 'mode partition functions' have already been obtained in the appropriate earlier sections. This is a factorization into mode partition functions, and the beauty of it is that, since all the thermodynamic functions depend on $\ln q$ rather than q directly, thermodynamic properties are additive. For example, the heat capacity is simply the sum of translational, rotational, vibrational, and electronic terms.

5.7 Two weakly interacting distinguishable molecules in a box

Suppose each molecule alone in the box is known to have $q = q_A$ or q_B. For instance A might be carbon dioxide, and B might be ammonia. Then, provided the interaction is vanishingly small, the partition function for the pair in the box must be

$$q = q_A q_B$$

The argument is the same as in the previous section, namely that, since each molecule has available states which are unaffected by which state the other happens to be in, the summation over energies available to the pair can be performed in two steps – first summing over states available to A and the over those available to B. Hence

$$q = \sum_{i,j} \exp\left[-\frac{\epsilon_{iA} + \epsilon_{jB}}{kT}\right]$$

$$= \left[\sum_i \exp\left(-\frac{\epsilon_{iA}}{kT}\right)\right]\left[\sum_j \exp\left(-\frac{\epsilon_{jB}}{kT}\right)\right]$$

$$= q_A q_B \quad (5.46)$$

This is a factorization into particle partition functions, and the argument can readily be extended to deal with two or more weakly interacting systems in thermal equilibrium, for instance a mixture of gases. We would expect

$$Q = Q_A Q_B Q_C \ldots \tag{5.47}$$

5.8 Conclusion

All the results of this chapter will prove of value when we have to deal with many-atom systems, but it must be remembered that the idea of using statistical mechanical techniques on single-particle systems should be applied with caution because energy fluctuations are not small in these systems. Thus, the requirements of sharp probability functions and steady observed properties are not fulfilled. However, the technique of factorizing Q into sub-system qs will prove enormously powerful in later chapters.

CHAPTER 5: SUMMARY OF MAIN POINTS

1. This chapter is devoted to the evaluation of the canonical partition functions for a number of elementary 'sub-systems'.
2. Although such sub-systems may show large fractional energy deviations, their partition functions q are essential algebraic building blocks for the later construction of partition functions Q for macroscopic systems.
3. Thermodynamic properties such as internal energy, heat capacity, pressure, and entropy are deduced from q in some cases, using the sub-system bridge equation $f = -kT \ln q$.
4. The cases considered are (5.2) a particle with only two possible energy states, (5.3) a simple harmonic oscillator, (5.4) a rotating heteronuclear diatomic molecule, (5.5) a particle in a box, (5.6) a heteronuclear molecule in a box, including rotation, vibration, and translation, and (5.7) two or more weakly interacting distinguishable molecules in a box.
5. Attention is drawn to cases where the third law of thermodynamics can be seen to be relevant, and to cases where results are reminiscent of those for macroscopic systems.

QUESTIONS

5.1 Consider a single atom in a box. The possible electron configurations of the atom have energies 0 (singly degenerate), ϵ (doubly degenerate) and 2ϵ (singly degenerate). Calculate the heat capacity and entropy of the single-atom system and sketch them as functions of temperature. Discuss the form of the entropy curve in terms of the uncertainty. Illustrate the artificiality of treating an atom as a system by obtaining an expression for the fractional fluctuation in energy.

5.2 Find expressions for the fractional deviation in energy of rotation of a single heteronuclear diatomic molecule in the low- and high-temperature limits.

5.3 The two halves of an ethane molecule can rotate about the C–C bond. How would you expect (qualitatively only) the potential energy of the molecule to depend on the angle of rotation? Discuss how you would expect the heat capacity of this mode of excitation to vary with T, (a) at sufficiently high T and (b) at sufficiently low T. Show why it is plausible that the heat capacity goes through a maximum between these two regions.

5.4 The carbon dioxide molecule (CO_2) is a linear molecule and it possesses four vibrational modes, with characteristic temperatures $\theta_v = 3360$ K, 1890 K, 954 K and 954 K again (i.e., there are two distinct modes with the same frequency). Calculate the vibrational heat capacity of the molecule at 312 K.

5.5 Calculate the fractional deviation in energy for two distinguishable particles in a box and compare this result with equation (5.42).

5.6 Write a computer program to evaluate and plot the heat capacity of a rotating heteronuclear diatomic molecule as a function of temperature. Show that it has a maximum at a temperature T_m which is approximately given by $T_m = 0.80\,\theta_r$. Also show that the heat capacity is within 1 % of the classical value k when $T \geqslant 1.94\,\theta_r$. [Hint: use the dimensionless variables c_V/k and T/θ_r. Since evaluating summations numerically is easier than peforming numerical differentiation, you may prefer to start with equation (5.24).]

5.7 Evaluate the partition function

$$q = \sum_{n=1}^{\infty} \exp\left(-n^2\theta_t/T\right)$$

numerically, for various values of T/θ_t. Find the lowest value of T/θ_t for which the approximate expression

$$q = (\pi T/4\theta_t)^{1/2}$$

differs from the exact expression by no more than 0.1 %.

CHAPTER 6
Models of Simple Solids

6.1 Introduction
6.2 The Einstein model
6.3 A one-dimensional oscillating lattice model
6.4 A three-dimensional oscillating lattic model
6.5 The Debye model
6.6 An alternative approach
Questions
References

6.1 Introduction

In this chapter we consider models of simple solids of N particles, where N is a large number so that the system may be considered macroscopic. Thus from the results of Chapter 4, we would expect the fluctuations to be vanishingly small, i.e., of order $N^{-1/2}$. Some of the results we obtained in the last chapter will prove to be very useful, for reasons we will come to in the next paragraph. First, however, we have to decide which distribution to use, the microcanonical, the canonical or the grand canonical. In Chapter 4 we treated the simplest realistic model of a solid, namely the Einstein solid, in the microcanonical distribution. In this chapter we shall use the canonical distribution, a choice dictated solely by mathematical convenience. Since we expect the fluctuations to be vanishingly small, this choice should not make any difference to the results. However, the reader will find hints in Section 6.6 of a possible alternative approach for one particular case and may wish to try applying the grand canonical distribution to the other models.

It will appear in a latter chapter that complications can arise in statistical mechanics from the fact that atomic particles are quantum-mechanically indistinguishable, that is, they cannot be labelled a, b, c, etc. By treating solids first we postpone that problem since, although atoms may be indistinguishable, lattice sites are not because they can each be labelled with a different lattice vector. Consequently we can take as a sub-system 'the atom on lattice site a', and if a canonical partition function for that site-atom can be evaluated then equation (5.47) surely applies, namely,

$$Q = q_a q_b q_c \ldots \quad (6.1)$$

This result is an important one, applying in general to assemblies of distinguishable non-interacting sub-systems. So far, it has been arrived at in a plausible way in Sections 5.6, 5.7 and 5.8. Later, in Chapter 8, a proof will be given that, for the special case where the basic entities are distinguishable (for example by their position in a lattice) but identical, then $Q = q^N$. A more general proof is given in Appendix 3. In this chapter we shall take equation (6.1) as a basis for further discussion of several possible models of the thermodynamic consequences of the vibrations of solid lattices.

6.2 The Einstein model

This model attempts to present a solid as an assembly of *non-interacting* site-atoms so that $q_a = q_b = q_c = \ldots = q$ and $Q = q^N$. Of course, atoms in a solid actually interact strongly, being strongly bound together, but in the Einstein model the following defining assumptions are made. Each atom is supposed to find itself in a potential well, due to all the other atoms in the solid (the principle contributions being from its nearest neighbours). Although all atoms are vibrating about their mean positions, it is assumed that, for the purposes of evaluating q for each site-atom, all *other* atoms are at rest at their mean positions. This is equivalent to taking the time average of the potential at the site-atom. Clearly the nature of the potential well is determined in general by the pair potential and the crystal structure, but the physics of the situation will be expressed more simply if we take the potential well to be harmonic and spherically symmetric. The Schrödinger equation for one oscillator has solutions for the allowed energy levels relative to rest-equilibrium energy, given by

$$\epsilon = \hbar\omega[(n_x + \tfrac{1}{2}) + (n_y + \tfrac{1}{2}) + (n_z + \tfrac{1}{2})] \quad (6.2)$$

where

$$n_x, n_y, n_z = 0, 1, 2, \ldots \quad (6.3)$$

and ω is the classical angular frequency. Since the three terms are independent, the result in (6.1) applies since the oscillator can be regarded as the combination of three independent one-dimensional SHOs. The result is that

MODELS OF SIMPLE SOLIDS

$$q = q_x q_y q_z \tag{6.4}$$

where

$$q_x = \sum_{n_x=0}^{\infty} \exp\left[-\frac{\hbar\omega(n_x + \tfrac{1}{2})}{kT}\right]$$

$$q_y = \sum_{n_y=0}^{\infty} \exp\left[-\frac{\hbar\omega(n_y + \tfrac{1}{2})}{kT}\right] \tag{6.5}$$

and

$$q_z = \sum_{n_z=0}^{\infty} \exp\left[-\frac{\hbar\omega(n_z + \tfrac{1}{2})}{kT}\right]$$

These expressions are all equal and have been evaluated in Section 5.3 dealing with the simple harmonic oscillator. Hence

$$q = \left[\tfrac{1}{2}\operatorname{cosech}\left(\frac{\theta_E}{2T}\right)\right]^3 \tag{6.6}$$

where $\theta_E \equiv \hbar\omega/k$ defines the characteristic Einstein temperature. Now, since all site-atoms have identical environments in this model, q is the same for them all. Thus, from equation (6.1),

$$Q = q^N = \left[\tfrac{1}{2}\operatorname{cosech}\left(\frac{\theta_E}{2T}\right)\right]^{3N} \tag{6.7}$$

The canonical bridge equation (4.37) immediately gives the Helmholtz free energy,

$$F = -kT\ln Q = +3NkT\ln\left[2\sinh\left(\frac{\theta_E}{2T}\right)\right] \tag{6.8}$$

The entropy and thermal capacity follow immediately:

$$S = -\left(\frac{\partial F}{\partial T}\right)_V = 3Nk\left[\left(\frac{\theta_E}{2T}\right)\coth\left(\frac{\theta_E}{2T}\right) - \ln\left\{2\sinh\left(\frac{\theta_E}{2T}\right)\right\}\right] \tag{6.9}$$

and

$$C_V = T\left(\frac{\partial S}{\partial T}\right)_V = 3Nk\left(\frac{\theta_E}{2T}\right)^2 \operatorname{cosech}^2\left(\frac{\theta_E}{2T}\right) \tag{6.10}$$

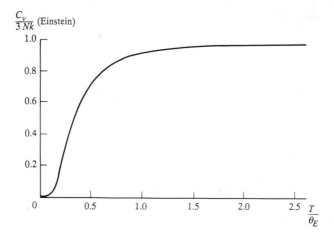

Figure 6.1 The theoretical Einstein heat capacity of a solid, equation (6.10).

The internal energy follows either from the definition of F, giving

$$U = F + TS \qquad (6.11)$$

or from equation (4.40). In either case the result is

$$U = \tfrac{3}{2} Nk\theta_E \coth\left(\frac{\theta_E}{2T}\right) \qquad (6.12)$$

Experimentally, the simplest quantity to measure is the specific heat. At high temperatures, experiment shows that the Dulong and Petit law is very well obeyed, a result also predicted from the pre-quantum idea of equipartition of energy. The theoretical Einstein heat capacity is shown in Figure 6.1. One can easily see from (6.10) that for $T \gg \theta_E$ (high temperatures) the result is $C_V = 3Nk$, as predicted by the Dulong and Petit law for N subsystems each with six degrees of freedom. However for $T \ll \theta_E$ (low temperatures) the Einstein model predicts

$$C_V = 3Nk \left(\frac{\theta_E}{T}\right)^2 \exp\left(-\frac{\theta_E}{T}\right) \qquad (6.13)$$

This is in contrast to the experimental results for most non-metallic solids. Figure 6.2 shows the experimental results for the heat capacity of solid argon over the temperature range 0 K to 2 K. The results show that the heat capacity is very accurately proportional to T^3. Making the Einstein potential well of a different shape or anisotropic will not produce a T^3 law. However, qualitatively the prediction is correct, in that $C_V \to 0$ as $T \to 0$

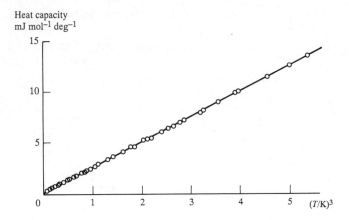

Figure 6.2 The experimental low-temperature heat capacity data for solid argon obtained and tabulated under 'Run I' by L. Finegold and N.E. Phillips, *Phys. Rev.* (1969) 177, 1383. The data is plotted versus $(T/K)^3$ to show how precisely the law is obeyed at temperatures.

and $C_V \to 3Nk$ as $T \to \infty$. In fact it is necessary experimentally to take very good low scatter data (as shown in Figure 6.2) to distinguish the T^3 behaviour from the behaviour predicted by equation (6.13).

The magnitude of the characteristic temperature θ_E is determined by the details of the interatomic potential and the interatomic spacing. To get an order of magnitude we can calculate the frequency of oscillation in a Lennard–Jones potential given by equation (1.1), at the equilibrium separation $2^{1/6}R$. The result is

$$\omega = \frac{6}{2^{1/6}} \left(\frac{\epsilon}{m\sigma^2}\right)^{1/2} = 2.48 \times 10^{12} \text{ rad s}^{-1} \qquad (6.14)$$

for argon. Hence $\theta_E = \hbar\omega/k = 19$ K, so that even at liquid nitrogen temperature the solid will obey the Dulong and Petit law. By contrast, however, strongly bonded crystalline diamond has $\theta_E = 132$ K; consequently there is a substantial reduction in heat capacity below the classical value even at room temperatures.

So far we have only calculated the heat capacity and we could in principle go on and calculate other thermodynamic quantities including the equation of state. The latter is obtained in the normal way by differentiating the free energy (6.8) with respect to the volume and equating it to minus the pressure. At first sight this would appear to give a zero pressure, since F is not explicitly dependent upon the volume. However, as we remarked above, the Einstein frequency and hence the characteristic temperature θ_E depend upon the atomic separation, which is proportional to $V^{-1/3}$. We will not carry out the explicit calculation since it is not warranted by the crudeness of the model.

Using equation (4.42) we can calculate the energy fluctuations directly as

$$\frac{\langle \Delta E \rangle_{\rm rms}}{U} = \frac{kT}{U}\left(\frac{C_V}{k}\right)^{1/2} = \left(\frac{1}{3N}\right)^{1/2} {\rm sech}\left(\frac{\theta_E}{2T}\right) \tag{6.15}$$

As $T \to \infty$, sech$(\theta_E/2T)$ approaches its maximum value of unity, so the fractional deviations are never greater than $(3N)^{-1/2}$, which shows that when N is macroscopic in magnitude, the fluctuations are entirely negligible. This confirms our general conclusions of Section 4.3 and the discussion in Chapter 1 that the probability distribution has to be sharply peaked. It also explains why the results obtained in this section are identical to those obtained in Section 4.2, treating the same model but using the microcanonical distribution. There the energy is specified (within a narrow range) rather than the mean energy.

We remarked at the end of Chapter 3 that our whole approach, based on the concept of uncertainty and its maximization, would be verified by the comparison of its predictions with experimental results. We should not, however, take the partial failure of the Einstein model too seriously, since its assumptions are manifestly crude. Its considerable degree of success leads us to hope that for an improved model, agreement would be even better. Clearly one assumption which is rather drastic, is the approximation that all atoms oscillate independently. The next section discusses a one-dimensional model, in which the atoms are coupled to their nearest neighbours, and this is a prelude to the more complex problem of coupled atoms in three dimensions.

6.3 A one-dimensional oscillating lattice model

Consider a chain of identical atoms, linked by 'springs' of Hooke's constant c and unstretched length a, as shown in Figure 6.3.

The atoms, of mass m, are restrained to move in directions parallel to the chain, and the equation of motion of each can be expressed in terms of positions of its neighbours. We shall further suppose that the ends of the chain (atoms $n = 0$ and $n = N$) are fixed, although other boundary conditions could be used without much affecting the results, provided that N is large. If x_n denotes the displacement of the nth atom from its position when the whole chain is at rest, then the classical equation of motion is

$$m\frac{d^2 x_n}{dt^2} = -c(x_n - x_{n-1}) - c(x_n - x_{n+1}) \tag{6.16}$$

MODELS OF SIMPLE SOLIDS

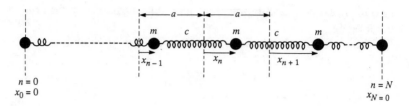

Figure 6.3 A one-dimensional oscillating lattice model. There are $(N-1)$ identical mobile particles of mass m and labelled $n = 1$ to $n = N - 1$. Particles labelled $n = 0$ and $n = N$ are held motionless. At rest, the particles are separated by a and connected by springs of Hooke's constant c. In motion (restricted to one dimension along the length of the chain) the displacement of the nth particle from its rest-equilibrium position is denoted by x_n.

Needless to say, we do not propose to carry out a purely classical analysis of the system, since it is evident from the Einstein model that the details of the quantum-mechanical allowed states and energies are of overriding importance, particularly at low temperatures. However, the quantum-mechanical approach requires a suitable form for the Hamiltonian, and this form is arrived at by pursuing equation (6.16) a little further. The possible solutions of this equation, appropriate to the chosen boundary conditions are

$$x_{nj}(t) = A_j \sin\left(\frac{j\pi n}{N}\right) \sin(\omega_j t + \varphi_j); \quad j = 1, 2, 3, \ldots (N-1) \quad (6.17)$$

where

$$\omega_j = \omega_m \sin(j\pi/2N) \quad (6.18)$$

and

$$\omega_m = 2(c/m)^{1/2} \quad (6.19)$$

A_j and φ_j denote the amplitude and the phase angle respectively. No further values of j are allowed since $j = 0$ and $j = N$ both describe motionless situations with $x_n = 0$ for all times. There are therefore $(N-1)$ modes each with a different label j, and there are also $(N-1)$ mobile atoms. Figure 6.4 shows a plot of ω_j/ω_m versus j/N, and each of the $(N-1)$ dots represents a mode of the chain.

Physically each mode is a standing wave of wavelength $\lambda = 2Na/j$, as is shown by the nature of the expression for x_{nj}. These waves are compression or sound waves and it is interesting to observe that their phase velocity is

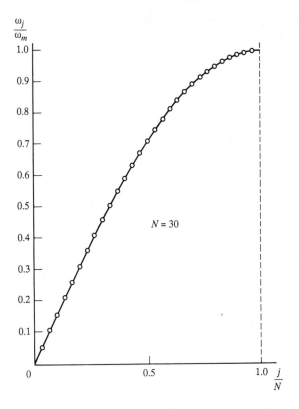

Figure 6.4 The allowed angular frequencies for $N = 30$ as given by equation (6.18). There are $(N - 1) \equiv 29$ allowed values. For higher values of N the shape of the plot would remain the same, but the discrete dots would become closer together until, as $N \to \infty$, a quasi-continuous line would result.

$$u = \tfrac{1}{2} a \omega_m \left[\frac{\sin\left(\dfrac{j\pi}{2N}\right)}{\left(\dfrac{j\pi}{2N}\right)} \right] \quad (6.20)$$

Thus the waves are dispersive unless $j \ll 2N/\pi$.

In general, the chain will not be oscillating in a pure mode, corresponding to one particular j-value, but rather a linear combination of modes. In considering this, let us notionally specify the instantaneous positions and velocities of all the $(N - 1)$ mobile atoms. Then subsequently the motion of the nth atom is described by summing x_{nj} over all possible modes j:

$$x_n(t) = \sum_{j=1}^{N-1} A_j \sin\left(\frac{j\pi n}{N}\right) \sin(\omega_j t + \varphi_j) \quad (6.21)$$

82   MODELS OF SIMPLE SOLIDS

The specification of the initial configuration of the chain is sufficient to determine all the A_j and φ_j, so the problem is solved. The next step is to calculate the total energy E where

$$E = E_k + E_p = \tfrac{1}{2}m \sum_{n=0}^{N-1} \dot{x}_n^2 + \tfrac{1}{2}c \sum_{n=0}^{N-1} (x_{n+1} - x_n)^2 \tag{6.22}$$

Substituting (6.21) and its time derivative into (6.22) (see Question 6.1) leads to the following expression for the total energy given by (6.22):

$$E = \tfrac{1}{2}m \sum_j (\dot{\xi}_j^2 + \omega_j^2 \xi_j^2) \tag{6.23}$$

where

$$\xi_j = \left(\frac{N}{2}\right)^{1/2} A_j \sin(\omega_j t + \varphi_j) \tag{6.24}$$

are known as the 'normal mode co-ordinates'. Differentiation with respect to t gives

$$\dot{\xi}_j = \left(\frac{N}{2}\right)^{1/2} \omega_j A_j \cos(\omega_j t + \varphi_j) \tag{6.25}$$

The significance and indeed elegance of equation (6.23) arises from the recognition of its character as the sum of the energies of $(N-1)$ classical harmonic oscillators, each distinguishable by its angular frequency ω_j given by equation (6.18). Quantum mechanically, therefore, the total Hamiltonian H is the sum of $(N-1)$ independent SHO Hamiltonians, each having energy eigenvalues

$$\epsilon_j = \hbar \omega_j (n_j + \tfrac{1}{2}) \tag{6.26}$$

where

$$n_j = 0, 1, 2, \ldots$$

The logic of this step, though plausible, is elaborated further in Chapter 8, but we adopt it here without further discussion. We thus have a situation very much like that of Section 5.6 relating to independent modes of a single molecule, and the consequence is that we should regard a *mode* of the oscillating chain as a *sub-system*. Each mode has an SHO canonical partition function; hence, using equation (5.15),

$$q_j = \tfrac{1}{2}\operatorname{cosech}\left(\frac{\hbar\omega_j}{2kT}\right) \tag{6.27}$$

and

$$Q = \prod_{j=1}^{N-1} q_j \tag{6.28}$$

At this point, we note that in the Einstein model Q factorized in terms of site-atom partition functions q, in which the frequency of each was *the same*. In the present case the range of frequencies complicates the mathematics. Thus we have the sum

$$\ln Q = -\sum_{j=1}^{N-1} \ln\left[2\sinh\left(\frac{\hbar\omega_j}{2kT}\right)\right] \tag{6.29}$$

where ω_j is given by equation (6.18). For macroscopic N the reader may easily verify that there are a vast number of nearly equal tiny terms in the sum so that transition to an integral is allowable as in Section 5.4 and elsewhere. This can be done by treating j as a quasi-continuous variable and, deriving from equation (6.18), that

$$dj = \frac{2N}{\pi}\frac{d\omega_j}{\omega_m \cos\left(\dfrac{j\pi}{2N}\right)} = \frac{2N}{\pi}\frac{d\omega_j}{(\omega_m^2 - \omega_j^2)^{\frac{1}{2}}} \tag{6.30}$$

This number dj is equal to the number of modes with frequencies within $d\omega_j$ at ω_j. Thus, dropping the subscript j,

$$\ln Q = \frac{-2N}{\pi}\int_0^{\omega_m}\frac{\ln\left[2\sinh\left(\dfrac{\hbar\omega}{2kT}\right)\right]d\omega}{(\omega_m^2 - \omega^2)^{\frac{1}{2}}} \tag{6.31}$$

Hence using the canonical bridge equation $F = kT\ln Q$, we find

$$S = -\left(\frac{\partial F}{\partial T}\right)_V = \frac{2Nk}{\pi}\int_0^{\omega_m}\frac{\dfrac{\hbar\omega}{2kT}\coth\left(\dfrac{\hbar\omega}{2kT}\right) - \ln\left[2\sinh\left(\dfrac{\hbar\omega}{2kT}\right)\right]}{(\omega_m^2 - \omega^2)^{\frac{1}{2}}}d\omega \tag{6.32}$$

and

$$C = T\left(\frac{\partial S}{\partial T}\right)_V = \frac{N\hbar^2}{2\pi kT^2} \int_0^{\omega_m} \frac{\omega^2 \operatorname{cosech}^2\left(\frac{\hbar\omega}{2kT}\right) d\omega}{(\omega_m^2 - \omega^2)^{\frac{1}{2}}} \qquad (6.33)$$

Although this expression can be evaluated by numerical integration, it is interesting and instructive to examine the nature of the integrand. At high temperatures ($T \gg \hbar\omega_m/k \geqslant \hbar\omega/k$),

$$\operatorname{cosech}\frac{\hbar\omega}{2kT} \to \left(\frac{2kT}{\hbar\omega}\right)$$

and

$$C = \frac{2N}{\pi} \int_0^{\omega_m} \frac{d\omega}{(\omega_m^2 - \omega^2)^{\frac{1}{2}}} \qquad (6.34)$$

$$= Nk \qquad (6.35)$$

This high-temperature result is exactly the same as the prediction of the one-dimensional Einstein model, although it is clear that in the present model a range of frequencies is important, rather than just one. The low-temperature limiting case can be found by observing that the numerator of the integrand in equation (6.33) has its maximum value of $(2kT/\hbar)^2$ for $\omega = 0$ at any temperature T, falling to below 1% of this maximum for $\omega/\omega_m = 10kT/\hbar\omega_m \ll 1$ in the low temperature limit. It follows that at low temperatures it is only important to know the low-frequency (i.e. long wavelength) modes accurately, since the rest do not contribute significantly. This observation forms the basis of the Debye model of a solid discussed in Section 6.5. Meanwhile, the integral for the low temperature heat capacity may be greatly simplified by ignoring ω in relation to ω_m, taking the upper limit of integration to be ∞, and substituting $x = \hbar\omega/2kT$. Then,

$$C = \frac{4Nk}{\pi}\frac{kT}{\hbar\omega_m}\int_0^\infty x^2 \operatorname{cosech}^2 x \, dx \qquad (6.36)$$

The integral is standard and equal to $\pi^2/6$, so

$$C = \frac{2}{3}\pi Nk \frac{kT}{\hbar\omega_m}, \quad \text{when} \quad T \ll \frac{\hbar\omega_m}{k} \qquad (6.37)$$

This low temperature result is certainly different from the prediction of the one-dimensional Einstein model with $T \ll \hbar\omega_E/k$, namely,

$$C(\text{Einstein } 1-D) = Nk \left(\frac{\hbar\omega_E}{kT}\right)^2 \exp\left(\frac{\hbar\omega_E}{kT}\right) \tag{6.38}$$

In this expression ω_E is the single Einstein frequency. However, one-dimensional cases are impossible to check against experiment, and our aim must now be to extend the lattice-dynamic model to three dimensions with the hope of improving on the Einstein model. We may hope that an understanding will emerge of why solid argon (see Figure 6.2) and other dielectric crystals have low-temperature heat capacities proportional to T^3.

6.4 A three-dimensional oscillating lattice model

We can take it[1] that the mode description works just as well in three dimensions as in one, provided as before that only harmonic forces are assumed. It follows that for each mode j there is an energy level scheme which is formally identical with that of a three-dimensional harmonic oscillator. In all there will be (for N mobile atoms) $3N$ modes or, if preferred, $3N$ one-dimensional simple harmonic oscillator frequencies ω_j. Thus, as in Section 6.3,

$$\ln Q = -\sum_{j=1}^{3N} \ln\left[2\sinh\left(\frac{\hbar\omega_m}{2kT}\right)\right] \tag{6.39}$$

Rather than attempt to carry out the nearly impossible task of evaluating the sum, we shall make use of the fact that the frequencies are very closely bunched together in the same way as they were for the one-dimensional chain. Thus, replacing the sum by an integral, i.e.,

$$\ln Q = -\int_0^\infty \ln\left[2\sinh\left(\frac{\hbar\omega}{2kT}\right)\right] g(\omega)\,d\omega \tag{6.40}$$

where $g(\omega)d\omega$, known as the frequency spectrum, is the number of modes (or oscillators) with angular frequency within $d\omega$ at ω. As for the one-dimensional lattice model of the previous section we must have

$$\int_0^\infty g(\omega)\,d\omega = 3N \tag{6.41}$$

which ensures that the total number of oscillators is $3N$. In general, the frequency spectrum depends on the details of the interatomic or intermolecular forces and the crystal structure. Figure 6.5 shows a detailed numerical computation of the frequency spectrum for the diamond lattice, which shows quite considerable structure.

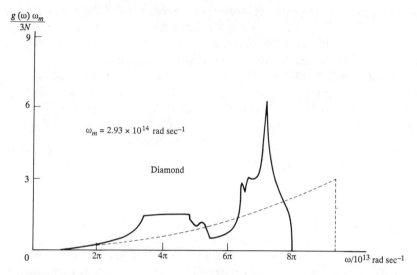

Figure 6.5 The frequency spectrum of diamond given by Dolling and Cowley, *Proc. Phys. Soc.* (1966) **88**, 463. For comparison, the Debye model approximation is given by the dashed curve with $\omega_m = 2.93 \times 10^{14}$ rad sec^{-1} as the cut-off frequency. There is not a strong resemblance between the curves (!) except that they both have sharp cut-off frequencies (though with different values), they coincide as $v \to 0$, and they have the same area. The dimensionless quantity $g(\omega)\omega_m/3N$ is plotted vertically.

It is not difficult to show that the only quantity we need to calculate all the thermodynamic properties is the frequency spectrum. Thus the bridge equation is

$$F = -kT \ln Q \tag{6.42}$$

Hence

$$S = -\left(\frac{\partial F}{\partial T}\right)_V = k \int_0^\infty \left\{ \left(\frac{\hbar\omega}{2kT}\right) \coth\left(\frac{\hbar\omega}{2kT}\right) \right.$$
$$\left. - \ln\left[2\sinh\left(\frac{\hbar}{2kT}\right)\right] \right\} g(\omega)\,d\omega \tag{6.43}$$

and

$$C_V = T\left(\frac{\partial S}{\partial T}\right)_V = k \int_0^\infty \left(\frac{\hbar\omega}{2kT}\right)^2 \operatorname{cosech}^2\left(\frac{\hbar\omega}{2kT}\right) g(\omega)\,d\omega \tag{6.44}$$

It should be noted that we have formally taken the upper limit of the integrals to be infinite, but if there is a finite cut-off as in the case of the diamond lattice, then we simply put the frequency spectrum equal to zero above the cutoff. The upper limit of the integrals then becomes the cut-off frequency. We can go on and calculate the equation of state, since $p = -(\partial F/\partial V)_T$. The only volume dependence of F is through $g(\omega)$ so that

$$p = -kT \int_0^\infty \ln\left[2\sinh\left(\frac{\hbar\omega}{2kT}\right)\right] \frac{\partial g(\omega)}{\partial V} d\omega \qquad (6.45)$$

In the next section we present a simple three-dimensional model that enables $g(\omega)$ to be calculated explicitly.

6.5 The Debye model

The aim is to devise a model close enough to reality to predict correctly low- and high-temperature heat capacities but idealized sufficiently to yield an analytic form for $g(\omega)$. The following guidelines are suggested by what has gone before:

(a) If there are N atoms, there must be $3N$ modes, having $q_i = \frac{1}{2}\text{cosech}(\hbar\omega_j/2kT)$.

(b) At low temperatures it is essential to have the right low-frequency limiting form of $g(\omega)$.

At high temperatures, all modes are fully thermally activated so their distribution is of little importance. Consequently, a simplified model which pays attention to (b) above is likely to be most in error at intermediate temperatures.

The importance of being right about the low frequency, i.e. long-wavelength form of $g(\omega)$ is easy to cope with, since long wavelengths are sure to be insensitive to the short periodicity of the lattice. Consequently, we may regard the lattice as a sort of continuous elastic jelly, whose modes can be determined by purely classical means. The wave equation for a small displacement u is

$$\frac{\partial^2 u}{\partial x^2} + \frac{\partial^2 u}{\partial y^2} + \frac{\partial^2 u}{\partial z^2} = \frac{1}{c^2}\frac{\partial^2 u}{\partial t^2} \qquad (6.46)$$

where c is the velocity of the wave.

Considering for convenience a cube of crystal of volume $V = L^3$ whose walls are held motionless, the appropriate solution is

MODELS OF SIMPLE SOLIDS

$$u = A \sin\left(n_x \frac{\pi x}{L}\right) \sin\left(n_y \frac{\pi y}{L}\right) \sin\left(n_z \frac{\pi z}{L}\right) \cos \omega t \tag{6.47}$$

where n_x, n_y and n_z are non-zero positive integers and

$$n_x^2 + n_y^2 + n_z^2 = \frac{L^2 \omega^2}{\pi^2 c^2} \tag{6.48}$$

This equation in fact describes a sphere of radius $L\omega/\pi c$ in (n_x, n_y, n_z)-space, in which unit volume represents one mode. Thus the number of modes with frequencies between zero and ω is simply the volume of an octant (because n_x, n_y, and n_z must all be positive) of the sphere, i.e.,

$$\frac{\pi L^3 \omega^3}{6 \pi^3 c^3} = \frac{V \omega^3}{6 \pi^2 c^3}$$

Hence, by differentiating and multiplying by three to take account of the fact that the wave-equation must describe a compressional displacement and two transverse displacements,

$$g(\omega) = \frac{3}{2} \frac{V \omega^2}{\pi^2 c^3} \tag{6.49}$$

In general, the velocities of the longitudinal and transverse waves are not equal, so we interpret c as an average defined by

$$\frac{3}{c^3} = \frac{1}{v_\varrho^3} + \frac{1}{v_{t_1}^3} + \frac{1}{v_{t_2}^3} \tag{6.50}$$

Now the frequency distribution (6.49) increases without limit as ω increases, so we have to impose a high frequency cut-off so that $g(\omega) = 0$ for $\omega > \omega_m$ in order to satisfy (6.41). That condition then becomes

$$\int_0^\infty g(\omega) d\omega = \int_0^{\omega_m} \frac{3 V \omega^2}{2 \pi^2 c^3} d\omega = 3N$$

Hence

$$\omega_m^3 = 6N \pi^2 c^3 / V \tag{6.51}$$

so that $g(\omega)$ can be written

and
$$g(\omega) = 9N\omega^2/\omega_m^3 \quad \text{for } \omega < \omega_m$$
$$g(\omega) = 0 \quad \text{for } \omega > \omega_m \qquad (6.52)$$

It is interesting to note that associated with ω_m there is a minimum wavelength given by

$$\lambda_{min} = 2\pi c/\omega_m = (4\pi V/3N)^{1/3} \qquad (6.53)$$

This is close to the value of the lattice spacing in the crystal and physically it is obvious that no wave can propagate whose wavelength is shorter than the lattice spacing. This provides a neat physical justification for the cut-off ω_m.

Since we have now calculated ω_m we can use equations (6.40) and (6.42) to (6.45) to calculate the thermodynamic properties. For the heat capacity we obtain

$$C_V = \frac{9N\hbar^2}{4\omega_m^3 kT^2} \int_0^{\omega_m} \omega^4 \text{cosech}^2\left(\frac{\hbar\omega}{2kT}\right) d\omega \qquad (6.54)$$

If we introduce the characteristic or Debye temperature

$$\theta_D = \hbar\omega_m/k \qquad (6.55)$$

$x = \hbar\omega/2kT$

and change the variable of integration to $x = \hbar\omega/kT$, (6.54) can be written

$$C_V = 72Nk(T/\theta_D)^3 \int_0^{\theta_D/2T} x^4 \text{cosech}^2 x \, dx \qquad (6.56)$$

Thus the heat capacities of all dieletric solids ought to be universal functions of T/θ_D. Question (6.13) asks the reader to calculate C_V over the whole temperature range. However, it is easy to evaluate it in the limiting cases of $T \gg \theta_D$ and $T \ll \theta_D$. In the first (high-temperature) case, the upper limit of the integral tends to zero. As $x \to 0$, cosech $x \to 1/x$, so the integrand approaches x^2; the integration can easily be performed to get the well known Dulong and Petit law. In the second (low-temperature) case, the upper limit of integration tends to infinity and the resulting integral is a standard one with the value $\pi^4/30$. Hence,

$$C_V = \frac{12Nk\pi^4}{5}\left(\frac{T}{\theta_D}\right)^3 \qquad (6.57)$$

This result agrees with experimental results for many dielectric solids and in particular the heat capacity for solid argon shown in Figure 6.2. This result

90 MODELS OF SIMPLE SOLIDS

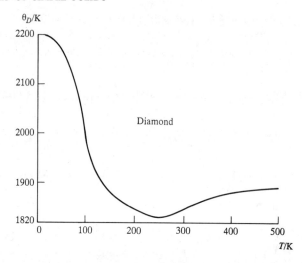

Figure 6.6 The behaviour of θ_D, used as a temperature-dependent parameter to fit the experimental heat capacity of diamond (NBS Monograph number 21 (1960), US Department of Commerce).

is extremely encouraging, in that it has enabled us to link the microscopic model with the macroscopic properties, the original task that we set ourselves in Chapter 1. To some extent we had achieved that with the Einstein model, which gave the correct high-temperature behaviour, but we now have also the correct low-temperature behaviour. However, at intermediate temperatures, the detailed structure of the frequency spectrum becomes important (see Figure 6.5 where the frequency spectrum of diamond is compared to the corresponding Debye spectrum). To see how this structure affects the heat capacity, the experimental results are sometimes analysed by fitting to a Debye expression but allowing the Debye temperature to vary weakly with temperature. The variation of θ_D is shown for diamond in Figure 6.6. Clearly there are significant differences from the Debye model, requiring a variation of θ_D over a range of about 20 % between $\theta_D = 1800\,K$ and $\theta_D = 2200$.

The calculation of the U, S and p is left as an exercise for the reader (see Question (6.14)). However one final point should be noted, that the energies derived from these expressions are referred to zero in the rest-equilibrium configuration. This will be of importance when we consider, in a later chapter, the question of phases in equilibrium.

6.6 An alternative approach

This chapter has concentrated entirely on the application of the canonical distribution. However, from equation (6.24) it is apparent that, apart from the 'zero-point energy' $\frac{1}{2}\hbar\omega_j$, we could have argued as follows. Solid vibrations may be viewed as an assembly of 'phonons', each having a characteristic angular frequency ω_j and associated energy $\hbar\omega_j$. We would then take the number n_j to denote the *number* of such phonons with ω_j. In such a view, the total number of phonons in the solid is not fixed so that the grand canonical distribution would be appropriate. Moreover, since the total number is a variable parameter, its magnitude will adjust itself to correspond to free energy minimization and this implies (cf. equation (12.7)) that $\mu = 0$. However, there is a significant and fundamental difference in this way of looking at things; the phonons are no longer distinguishable. The reader is not in a position to deal with this before reading Chapter 10. In Chapter 11 there is a similar treatment of electromagnetic waves where they are regarded as a gas of photons in the grand canonical distribution with $\mu = 0$. In practice, the approach we have adopted in this chapter is the easier. The phonon approach is probably most helpful when dealing with heat transport in solids, a topic beyond the scope of this book. These two different approaches do emphasize that alternatives are possible in statistical mechanics. In the next chapter this is illustrated in detail for a particular case.

CHAPTER 6: SUMMARY OF MAIN POINTS

1. The Einstein model of a solid treats all the atoms as moving in the same harmonic potential well. Application of the canonical distribution shows that the model can reproduce the classical high temperature behaviour. At low temperatures, it predicts the specific heat to have an exponential temperature dependence, whereas experimentally the specific heat is found to be proportional to T^3.
2. A one-dimensional chain of atoms interacting with a Hooke's law force can be solved exactly, enabling construction of the canonical partition function. The result is a specific heat proportional to the temperature.
3. The same model does not have an exact solution in three dimensions. In the Debye model the solid is replaced by an elastic continuum. This predicts the correct behaviour at both low and high temperatures but tends to be inaccurate in the intermediate temperature range.
4. In all the oscillating solid models, the apropriate quantity that has to be calculated is the frequency spectrum, that is, the number of modes of vibration in an infinitesimal frequency range.

QUESTIONS

6.1 By differentiating equation (6.21) with respect to time and substituting into equation (6.22) show that the kinetic energy of the linear chain can be written in the form

$$E_k = \tfrac{1}{2}mN\sum_j (\tfrac{1}{2}A_j)^2 \omega_j^2 \cos^2(\omega_j t + \varphi_j)$$

[Hint: the expression for E_k will involve three summations, over j, j' and n. Perform the sum over n first and show that it is equal to $\tfrac{1}{2}N\delta_{jj'}$.]

Follow a similar procedure to show that the potential energy is given by

$$E_p = \tfrac{1}{2}N\sum_j (\tfrac{1}{2}A_j)^2 \sin^2(\omega_j t + \varphi_j) 4c \sin^2\left(\frac{\pi j}{N}\right)$$

Hence obtain equation (6.23).

6.2 Some crystalline solids have a highly anisotropic layered structure. The flaky mica crystal is a good example of this. Calculate the room temperature heat capacity of such a solid, adopting the following model. Suppose the restoring forces on an atom in directions parallel to a layer are so large that the angular frequency of the motion in those two directions ω_1 obeys $\hbar\omega_1 \gg 300k$ (the thermal energy kT at room temperature). However, let the angular frequency ω_2 for motion perpendicular to the layers be such that $\hbar\omega_2 \ll 300k$.

6.3 Show that for the Einstein model of a solid, the fractional fluctuation in energy is

$$\frac{\langle \Delta E \rangle_{rms}}{U} = \left(\frac{1}{3N}\right)^{\frac{1}{2}} \mathrm{sech}\left(\frac{\hbar\omega}{2kT}\right)$$

At what temperature does this expression reach its greatest value, and what is that value?

6.4 Use a method analogous to the Debye theory to find the heat capacity associated with surface waves in a liquid in the low-temperature limit. You may take the relationship between frequency ν and wavelength λ to be $\nu^2 = 2\pi\sigma/\rho\lambda^3$, where σ is the surface tension and ρ the density. The surface tension of low-temperature liquid helium is 0.352 mN m^{-1} and its density is 145 kg m^{-3}. From this data estimate the temperature range over which your formula is valid for liquid helium, assuming that each helium atom at the surface possesses one degree of freedom. You may use the result

$$\int_0^\infty x^{7/3} \operatorname{cosech}^2 x \, dx = 1.5602$$

6.5 Obtain general expressions for the heat capacities of a one-dimensional (length L) and a two-dimensional (area $A = L^2$) Debye solid. Do this by calculating in each case the density of modes $g(\omega)$ and following the recipe for the three-dimensional case. Prove that in the n-dimensional case (with $n = 1, 2,$ or 3) the high-temperature results are nNk and that the low-temperature results are proportional to T^n. For the one-dimensional low-temperature case, compare and contrast the result with those obtained in this chapter for the one-dimensional Einstein and lattice-dynamic models.

6.6 Consider a solid in which there are $2N$ harmonic oscillators of frequency ν, $3N$ of frequency 2ν, and $4N$ of frequency 3ν. Obtain expressions for internal energy U and heat capacity C_V as functions of temperature. Compare your results with tabulated Debye functions (Question 6.13) to show how the characteristic Debye temperature θ_D of our model solid varies with temperature.

6.7 Heat capacity measurements for copper at $T = 115$ K yield 16.50 Jmol^{-1}K^{-1}. Using this value, calculate the values of the Einstein and Debye characteristic temperatures for copper and use them (with the appropriate expressions) to predict values of the heat capacity at 35 K. Compare results of these two calculations with the measured value for copper, i.e., 1.89 Jmol^{-1}K^{-1}. Which model (Einstein or Debye) is the more successful in this test?

6.8 According to the Dulong–Petit law, the molar heat capacity of all solids should be a temperature-independent constant equal to 24.9 Jmol^{-1}K^{-1}. In fact, at 440 K lead has 24.8 Jmol^{-1}K^{-1} and diamond has 9.2 Jmol^{-1}K^{-1}. Explain the discrepancy and calculate the temperature at which diamond has 95% of the Dulong–Petit value. You may take $\theta_D = 88$ K for lead and $\theta_D = 2200$ K for diamond, where the latter approximate value will be found to simplify the arithmetic.

6.9 Calculate the low-temperature heat capacity of a one-dimensional Debye solid and compare the result with that of the lattice-dynamic model.

6.10 Estimate the range of fractional energy deviations in a Debye solid.

6.11 In the Debye model of lattice vibrations, the number of longitudinal wave modes with wavelength within $d\lambda$ at λ is

$$\frac{4\pi V}{\lambda^4} d\lambda \quad \text{for } \lambda > \left(\frac{4\pi V}{3N}\right)^{1/3} \quad \text{(given)}$$

and

$$0 \quad \text{for } \lambda < \left(\frac{4\pi V}{3N}\right)^{1/3} \quad \text{(given)}$$

A mode of given wavelength λ has a corresponding frequency ν and it is a part of the model to assume that each mode has a heat capacity of the simple oscillator form:

$$C_{mode} = k\left(\frac{h\nu}{2kT}\right)^2 \text{cosech}^2\left(\frac{h\nu}{2kT}\right) \quad \text{(given)}$$

Usually these ideas are applied to the calculation of heat capacity due to quantized lattice wave for which $\nu\lambda = c$ (the velocity of sound). However, in a ferromagnetic solid at low temperatures, quantised waves of magnetization (spin waves) have $\nu\lambda^2 = A$ where A is a constant. At low temperatures, prove that the temperature dependence of the heat capacity due to such spin waves is proportional to $T^{3/2}$.

6.12 Consider a rigid lattice (constant volume V) of N (constant) atoms each of which has only two non-degenerate electronic states with energies ϵ and $\epsilon + \Delta$. Convince yourself that the canonical partition function is

$$Q = \exp\left(-\frac{N\epsilon}{kT}\right)\left(1 + \exp\left[-\frac{\Delta}{kT}\right]\right)^N$$

and derive an expression for the entropy. Does your expression satisfy the third law of thermodynamics? If not, why not?

6.13 Evaluate the integral in equation (6.56) numerically, for a range of values of T/θ_D. Hence plot the heat capacity as a function of T/θ_D. On the same curve plot the experimental values of the heat capacities for solid argon and for indium whose data is given below. (For the purposes of this question the differences between C_p and C_V need not be considered. Generally the difference between C_V and C_p is extremely small. See Question 6.14 below.)

ARGON

Note: solid argon is an insulator and the total measured specific heat is due entirely to lattice vibrations. The value of θ_D as $T \to 0$ can be deduced most easily from the low-temperature T^3 limiting form given in equation (6.57). The authors give the value as 92.0 K but you may wish to check that. The keen student may go on to plot $\theta_D(T)$.

The data labelled 'Run I' and 'Run II' are by L. Finegold and N.E. Phillips, published in *Phys. Rev.* (1969) **177**, 1383. The higher-temperature data labelled 'Series I', 'Series II' are by P. Flubacher, A.J. Leadbetter and J.A. Morrison, published in *Proc. Phys. Soc.* (1961). **78**, 1449.

Temp (K)	C (mJ mol^{-1} K^{-1})	Temp (K)	C (mJ mol^{-1} K^{-1})
		Run I	
4.8913	354.0	1.8332	15.71
5.4764	510.9	1.9575	19.11
6.0978	731.8	2.1398	25.11
6.6429	971.6	2.3312	32.73
7.1685	1235.0	2.5233	41.68
8.0024	1735.0	2.7478	54.33
8.7131	2216.0	2.9965	71.17
4.4551	255.5	3.2243	89.91
0.4978	0.3083	3.4492	111.1
0.5334	0.3741	3.7407	143.8
0.5693	0.4690	4.0687	189.9
0.5960	0.5364	4.4499	255.7
0.6292	0.6176	4.8235	331.1
0.6587	0.7065	1.4206	7.212
0.6896	0.8131	1.5774	9.906
0.8283	1.423	1.6606	11.56
0.8817	1.725	1.7543	13.71
0.9578	2.237	1.8423	15.89
1.0376	2.826	1.9193	17.85
1.1253	3.584	1.9920	20.21
1.1735	4.020	2.0690	22.71
1.2253	4.603	0.5104	0.3368
1.2844	5.288	0.5576	0.4270
1.3414	6.050	0.6003	0.5313
1.4059	6.973	0.6052	0.5487
1.4735	7.958	0.6613	0.7315
1.5289	9.035	0.7147	0.9246
2.0727	22.85	0.7621	1.130
2.1614	25.94	0.8100	1.317
2.1183	24.42	0.8617	1.616
2.2907	31.12	0.9265	2.007
2.3979	35.85	0.9896	2.439
2.5257	41.72	0.9508	2.173
2.6255	47.21	1.0195	2.648
2.7569	55.02	1.0927	3.293
2.9071	64.64	1.2683	5.144

MODELS OF SIMPLE SOLIDS

Temp (K)	C (mJ mol^{-1} K^{-1})	Temp (K)	C (mJ mol^{-1} K^{-1})
3.0710	76.88	1.3650	6.403
3.2732	94.38	4.3091	231.2
3.5483	121.5	4.8241	334.1
3.8604	160.2	5.3373	466.8
4.2258	216.1	5.8326	630.2
1.2105	4.507	6.3069	816.2
1.2989	5.480	6.9139	1103.0
1.3832	6.630	8.2931	1948.0
1.4828	8.167	9.0860	2480.0
1.5852	10.07	11.2261	4366.0
1.7119	12.77		

Run II

Temp (K)	C (mJ mol^{-1} K^{-1})	Temp (K)	C (mJ mol^{-1} K^{-1})
0.4367	0.2130	10.0900	3341.0
0.4840	0.2818	11.2269	4376.0
0.5332	0.3765	4.5362	275.1
0.5833	0.4976	4.9792	372.3
0.6374	0.6486	5.4175	495.7
0.6645	0.7397	5.9462	672.1
0.7100	0.8876	6.5000	906.8
0.7669	1.122	1.2380	4.785
0.8464	1.540	1.3722	6.523
0.8934	1.793	1.5863	10.13
0.9205	1.949	1.6516	11.42
1.0026	2.555	1.7380	13.42
1.1060	3.418	1.8316	15.75
1.2146	4.513	1.9309	18.43
1.2511	4.930	2.0162	21.19
1.3693	6.455	2.1260	24.72
1.4809	8.167	2.2769	30.53
0.3960	0.1553	2.4627	38.88
0.5527	0.4374	2.6562	49.03
0.6546	0.7025	2.8708	62.44
0.7183	0.9353	3.0846	78.30
0.7722	1.153	3.3288	99.79
0.8648	1.642	3.5949	127.9
0.9569	2.187	3.9063	167.0
1.0338	2.761	4.2602	222.1
1.1024	3.313	4.6114	289.6
1.1528	3.840	4.9798	375.1
1.2036	4.395	1.2428	4.838
1.3098	5.669	1.3559	6.299
1.3968	6.889	1.4604	7.863
1.4903	8.342	1.5269	9.020
0.4969	0.3041	1.6035	10.49
0.5910	0.5338	1.9228	18.24
0.6450	0.6813	2.0508	22.20

Temp (K)	C (mJ mol^{-1} K^{-1})	Temp (K)	C (mJ mol^{-1} K^{-1})
0.7042	0.8869	2.2399	29.21
0.7693	1.162	2.4571	38.78
0.8449	1.543	2.6673	49.92
4.1810	210.1	2.9187	66.09
4.6041	287.6	3.1743	86.27
5.1224	408.3	3.4365	111.5
5.6733	574.4	3.7313	144.6
6.2357	784.6	4.6987	310.2
7.5256	1456.0	5.1263	417.5
8.3630	1995.0	1.4946	8.44
9.1864	2598.0		

Temp (K)	C (J mol^{-1} K^{-1})	Temp (K)	C (J mol^{-1} K^{-1})
Series I		**Series II**	
17.201	10.09	2.220	0.02788
18.304	11.08	2.540	0.04123
19.504	12.07	2.693	0.05086
20.771	13.14	3.002	0.06798
22.086	14.12	3.205	0.09021
23.692	15.25	3.467	0.1126
25.694	16.40	3.686	0.1382
26.834	17.06	3.931	0.1662
27.053	17.18	4.191	0.1983
27.900	17.64	4.388	0.2405
28.792	18.04	4.690	0.3025
29.245	18.25	5.158	0.4117
30.467	19.11	5.674	0.5709
30.759	19.07	6.234	0.7932
31.233	19.25	6.761	1.041
32.557	19.86	7.310	1.351
33.247	20.15	7.918	1.722
34.446	20.64	8.826	2.365
35.570	21.05	9.877	3.205
36.634	21.44	10.818	4.023
37.978	21.91	11.333	4.508
38.937	22.23	11.839	4.972
40.455	22.63	12.374	5.487
41.225	22.89	12.900	6.019
42.880	23.32	13.323	6.442
45.147	23.86	13.837	7.040
47.219	24.42	14.309	7.388
49.274	24.95	14.806	7.957
51.385	25.49	15.314	8.357
53.403	25.95	15.820	8.907

Temp (K)	C (J mol^{-1} K^{-1})		Temp (K)	C (J mol^{-1} K^{-1})
55.472	26.45		16.389	9.397
57.602	26.67		17.514	10.41
59.635	27.29		18.727	11.44
61.596	27.77		20.063	12.59
63.491	28.01		21.431	13.68
65.435	28.54			
67.502	28.98			
67.529	28.69			
70.971	30.16			
72.736	30.78			
74.525	31.18			
76.333	31.95			
78.432	32.58			
78.931	32.55			
80.254	33.39			
81.120	33.86			
81.552	33.97			
82.970	34.66			
83.114	35.09			

INDIUM

Note: solid indium is a conductor and the total measured specific heat is due to lattice vibrations, *plus* a contribution from the free electron gas as discussed later in Section 11.2. The data tabulated below is for the total specific heat, as reported by J.R. Clement and E.H. Quinnell in *Phys. Rev.* (1953) **92**, 258. The authors estimated that the electron contribution was given by $1.800T$ mJ mol^{-1}K^{-1} (with T in degrees Kelvin). This quantity should therefore be first subtracted from the data below before fitting to the Debye expression. The value of θ_D as $T \to 0$ can be deduced most easily from the low-temperature T^3 limiting form given in equation (6.57). The authors give the value as 109 K but you may wish to check that. The keen student may go on to plot $\theta_D(T)$.

Temp (K)	C (mJ mol^{-1} K^{-1})	Temp (K)	C (mJ mol^{-1} K^{-1})
1.7	10.53	6.5	527.9
1.8	12.11	7.0	661.8
1.9	13.86	7.5	813.8
2.0	15.79	8.0	981.6
2.1	17.91	8.5	1164

Temp (K)	C (mJ mol^{-1} K^{-1})	Temp (K)	C (mJ mol^{-1} K^{-1})
2.2	20.23	9.0	1358
2.3	22.74	9.5	1565
2.4	25.48	10.0	1780
2.5	28.45	11.0	2231
2.6	31.66	12.0	2704
2.7	34.99	13.0	3194
2.8	38.84	14.0	3699
2.9	42.82	15.0	4215
3.0	47.09	16.0	4747
3.1	51.66	17.0	5287
3.2	56.55	18.0	5839
3.3	61.74	19.0	6400
3.4	67.06	20.0	6974
3.6	79.49	21.0	7560
3.8	93.47		
4.0	109.5		
4.5	161.6		
5.0	230.3		
5.5	313.1		
6.0	411.9		

6.14 Use equations (6.40), (6.42) and (6.52) to evaluate F in both the high and low temperature approximations. Hence evaluate S, U and p in the same limits. Show that these expressions are consistent with the statement made in Section 2.2 about the relation between p and U. Why is p not zero at $T = 0$? Derive the difference between the heat capacities at constant volume and the heat capacity at constant pressure and show that at least at low temperatures, the statement made in Question 6.13 above is correct.

(Hints: remember that the volume dependence of F comes through ω_m. Also you will need the following limiting approximations,

$$\mathop{\text{Lt}}_{x \to 0} \sinh x = 0 \qquad \mathop{\text{Lt}}_{x \to \infty} \sinh x = \tfrac{1}{2} e^x \qquad \mathop{\text{Lt}}_{x \to 0} x^n \ln x = 0$$

where n is a positive integer.)

REFERENCE

Goldstein H. (1980). *Classical Mechanics*, 2 edn. Addison-Wesley.

CHAPTER 7
Models of Magnetic Crystals

7.1 Introduction
7.2 Magnetic systems in the canonical distribution
7.3 Magnetic systems in the grand canonical distribution
7.4 Paramagnetic crystal in the canonical distribution
7.5 Paramagnetic crystal in the grand canonical distribution
Questions

7.1 Introduction

In Chapter 6 we discussed some simple harmonic models for the thermal properties of solids, making use of our intuitive argument that the canonical partition function could be factorized into the partition functions of simple sub-systems (for example, oscillators or normal modes). Many solids have interesting properties apart from those arising from lattice vibrations, not least of which are the magnetic properties. In this chapter we shall be concerned with the class of materials known as paramagnetic solids. These materials are insulators which have no magnetic moment in the absence of a field B, and for low fields have a magnetic moment proportional to B.

As a preparation for the general theory of distinguishable sub-systems given in Chapter 8, we shall not immediately factorize the partition function but start from the general partition function and carry out the sum explicitly. The result which we should expect is that, since the model we use is composed of independent distinguishable sub-systems, the partition function does factorize. We shall do this in both the canonical and grand canonical distributions and find that, as we would expect from our discussion of deviations in $\mathcal{N}$ in Section 4.4, the results are identical to order $1/\mathcal{N}$.

Before setting up the model, however, we must consider the link between the microscopic magnetic properties of a system and the macroscopic magnetic properties.

7.2 Magnetic systems in the canonical distribution

So far, for simplicity, we have confined ourselves to systems in which the macroscopic state has been specified by the thermodynamic variables U and V or alternatively T and V. However, for some systems we may have to specify other variables to completely define the macroscopic state. In particular for magnetic systems we would have to specify the magnetic field acting on the system. Similarly, for systems which are either electrically polarizable or possess a permanent electric dipole moment, the complete macroscopic state will not be specified unless we give the electric field acting on the sample. In this section we confine ourselves to magnetic systems. For systems in a magnetic field B, the first law of thermodynamics is modified to read

$$\delta Q = \delta U + p\delta V + M\delta B \tag{7.1}$$

for fixed number of particles N, where M is the total magnetic moment of the system. For such systems we are frequently interested in either the mechanical properties in zero field or the magnetic properties at constant volume. The case of zero field has occupied us up to now; we shall now consider the magnetic properties at constant volume. For this case, equation (7.1) becomes

$$\delta Q = \delta U + M\delta B \tag{7.2}$$

The second and third laws of thermodynamics are unchanged, namely

$$\delta Q = T\delta S \tag{7.3}$$

$$S \to 0 \text{ as } T \to 0 \tag{7.4}$$

Examination of equation (7.2) shows that we can obtain for the canonical ensemble the relevant formulae by making the substitutions

$$V \to B \text{ and } p \to M \tag{7.5}$$

Thus we obtain

$$F = -kT\ln Q \tag{7.6}$$

$$M = -\left(\frac{\partial F}{\partial B}\right)_T \tag{7.7}$$

and

$$S = -\left(\frac{\partial F}{\partial T}\right)_B \tag{7.8}$$

where

$$Q = \sum_n \sum_{\{j\}} \exp\left[-\frac{E_n(B)}{kt}\right] \tag{7.9}$$

and the energies E_n are to be regarded as functions of the field B only, the volume being a constant parameter in the energy. The reader can verify that this procedure is correct by following the analogous arguments to the mechanical case.

Our argument above would indicate that we could follow the same procedure for the grand canonical ensemble, but in fact there is a complication, which is not generally recognized. We therefore give the derivation of the bridge equations in more detail.

7.3 Magnetic systems in the grand canonical distribution

The argument given in Chapter 4 needs to be adapted to magnetic systems. It holds as far as the definition of the thermodynamic potential Z and the bridge equation (4.69). Using the relation

$$\mu\delta\mathcal{N} + T\delta S = \delta U + M\delta B \tag{7.10}$$

and differentiating Z, we find

$$\delta Z = \mathcal{N}\delta\mu + S\delta T + M\delta B \tag{7.11}$$

Thus

$$\mathcal{N} = \left(\frac{\partial Z}{\partial \mu}\right)_{T,B} \tag{7.12}$$

$$S = \left(\frac{\partial Z}{\partial T}\right)_{\mu,B} \tag{7.13}$$

and

$$M = \left(\frac{\partial Z}{\partial B}\right)_{\mu, T} \tag{7.14}$$

We could have obtained these results from equations (4.70) to (4.72) by using the substitution (7.5) above. However, the result $G = \mu \mathcal{N}$ is not valid and we cannot use the analogue to equation (4.75). This can be seen as follows. Using the substitution (7.5), the definition of G becomes

$$G = U + MB - TS \tag{7.15}$$

Differentiating this function and using the same procedure as for Z above we find

$$\delta G = B\delta M + \mu \delta \mathcal{N} - S\delta T \tag{7.16}$$

From this we conclude that

$$\left(\frac{\partial G}{\partial \mathcal{N}}\right)_{M,T} = \mu(M,T) \tag{7.17}$$

and hence

$$G = \mathcal{N}\mu(M,T) + f(M,T) \tag{7.18}$$

where f is an artitrary function of M and T. For the non-magnetic case we could eliminate f by using the fact that $G \to 0$ for $\mathcal{N} \to 0$, since in that case f was a function of p and T which are intensive variables. Here, however, M is extensive (proportional to $\mathcal{N}$) and thus, if we use this boundary condition, all we can say is that $f(0, T) = 0$. Hence in general $G \neq \mathcal{N}\mu$ and the analogy to equation (4.75) does not hold.

We shall now apply these ideas to the simple but interesting example of a paramagnetic crystal, using first the canonical distribution. Then, in the succeeding section the same model will be analysed using the grand canonical distribution to show the equivalence of the results.

7.4 Paramagnetic crystal in the canonical distribution

We suppose that the magnetic atoms of the lattice can be regarded as spins possessing a magnetic dipole moment. Now, according to the laws of quantum mechanics, if the total spin quantum number is S then the z component of the spin can take on the values, $S, S - 1, \ldots, -S$; furthermore S has to be integer or half integer. To simplify the mathematics, we shall treat the case of $S = \frac{1}{2}$. The N spins are taken to be localized in space and sufficiently far apart, on the average, that we can ignore interactions between them. In a magnetic field a single spin ($S = \frac{1}{2}$) has energy

$$\epsilon_\uparrow = -\tfrac{1}{2}\mu_B g B = -\gamma B$$
$$\epsilon_\downarrow = +\tfrac{1}{2}\mu_B g B = +\gamma B \qquad (7.19)$$

where μ_B is the Bohr magneton and g is the gyro-magnetic ratio. The energy level $\epsilon_\uparrow$ corresponds to a spin parallel to the field B, assumed in the z direction, and $\epsilon_\downarrow$ to a spin anti-parallel to the field. Both of these levels are non-degenerate. The energy levels of the total system are

$$E_N = N_\uparrow \epsilon_\uparrow + N_\downarrow \epsilon_\downarrow$$
$$= N_\uparrow(-\gamma B) + (N - N_\uparrow)\gamma B \qquad (7.20)$$

where $N_\uparrow$ and $N_\downarrow$ ($N_\uparrow + N_\downarrow = N$) are the total number of spins parallel and anti-parallel to the field. However, note that the complete state of the system is not specified by $N_\uparrow, N_\downarrow$. This is because we can rearrange the N spins on different lattice sites. To specify the quantum-mechanical state we have to specify the z component of the spin on every site. This set of numbers corresponds in our general notation to $\{j\}$, the energy quantum number $N_\uparrow$ corresponding to n. Since the total energy depends only on $N_\uparrow$ there will be a high degree of degeneracy. The number of degenerate states will be given by the number of ways we can choose $N_\uparrow$ spins from N spins, i.e.,

$$g_{N_\uparrow} = \frac{N!}{(N - N_\uparrow)! N_\uparrow!} \qquad (7.21)$$

We illustrate in Figure 7.1 how this degeneracy arises.

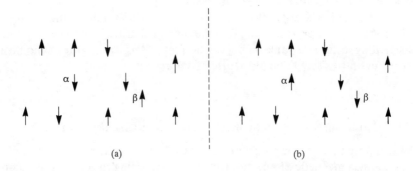

Figure 7.1 (a) and (b) are two possible states of a 'paramagnetic salt' of 11 spins. In both cases $N_\uparrow = 7$ and the energy of each is $-3\gamma B$. However, they are different states since in (b) the direction of two of the spins labelled α and β are reversed compared with (a). α and β are distinguishable since they are at different fixed points in space.

Note that from equation (7.20) the ground state is given by $N_\uparrow = N$ for which $g_{N_\downarrow} = 1$, which is consistent with the third law of thermodynamics. Using the definition of the canonical partition function (4.31) for this model, we have

$$Q = \sum_{N_\uparrow = 0}^{N} \frac{N!}{(N - N_\uparrow)! N_\uparrow!} \exp\left[-\frac{-N_\uparrow \gamma B + (N - N_\uparrow) \gamma B}{kT}\right]$$

$$= \sum_{N_\uparrow = 0}^{N} \frac{N!}{(N - N_\uparrow)! N_\uparrow!} \left[\exp\left(\frac{\gamma B}{kT}\right)\right]^{N_\uparrow} \left[\exp\left(-\frac{\gamma B}{kT}\right)\right]^{N - N_\uparrow}$$

Making use of the binomial theorem

$$(x + y)^N = \sum_{M=0}^{N} \frac{N!}{(N - M)! M!} x^M y^{N - M}$$

leads to

$$Q = \left[\exp\left(\frac{\theta_M}{T}\right) + \exp\left(-\frac{\theta_M}{T}\right)\right]^N$$

$$= \left[2\cosh\left(\frac{\theta_M}{T}\right)\right]^N \tag{7.22}$$

where we have defined the characteristic magnetic temperature

$$\theta_M = \gamma B / k \tag{7.23}$$

The result (7.22) is what we would have expected from our earlier intuitive argument, namely that for distinguishable identical sub-systems, $Q = q^N$. Thus

$$F = -NkT \ln\left[2\cosh\left(\frac{\theta_M}{T}\right)\right] \tag{7.24}$$

Also, using (7.8)

$$S = Nk\left\{\ln\left[2\cosh\left(\frac{\theta_M}{T}\right)\right] - \left(\frac{\theta_M}{T}\right)\tanh\left(\frac{\theta_M}{T}\right)\right\} \tag{7.25}$$

and, from (7.7)

$$M = N\gamma \tanh\left(\frac{\theta_M}{T}\right) \tag{7.26}$$

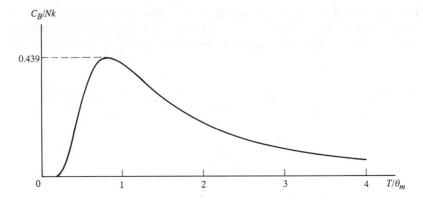

Figure 7.2 The heat capacity of a paramagnetic solid as a function of temperature.

We can also note that since $U = F + TS$,

$$U = -BM \tag{7.27}$$

which is the correct relationship between the magnetic energy and the magnetic moment. Finally, the heat capacity at constant magnetic field is

$$C_B = Nk \left(\frac{\theta_M}{T}\right)^2 \text{sech}^2 \left(\frac{\theta_M}{T}\right) \tag{7.28}$$

The heat capacity is plotted in Figure 7.2, and it has a maximum value of about $0.439Nk$ when $(T/\theta_M) \approx 0.834$. In Figure 7.3 the entropy and magnetization are plotted as functions of T for fixed θ_M, or equivalently

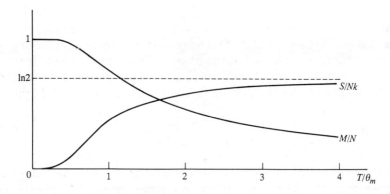

Figure 7.3 The entropy and magnetization of a paramagnetic solid as a function of temperature.

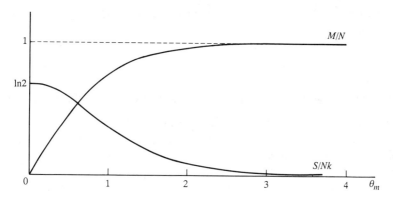

Figure 7.4 The entropy and magnetization of a paramagnetic solid as a function of magnetic field.

fixed B. In Figure 7.4 the same quantities are plotted as functions of θ_M at a fixed temperature. Figure 7.3 shows the usual result that as the temperature increases the entropy increases and consequently the uncertainty of the state of the system. The more uncertain the system the more will be its state of disorder, since there are many fewer ordered states than disordered ones, consequently the magnetization declines, since a high magnetization implies a high degree of ordered dipoles. Conversely, Figure 7.4 shows that as we increase the field, the magnetization increases, since the dipoles tend to line up with the field. The state of the system becomes more ordered and, consequently, the entropy decreases. This is a nice simple interpretation of our basic thesis of the relationship between entropy and uncertainty.

We can make these general arguments more specific. In the low temperature limit, (7.25) and (7.26) give

$$S \to 0 \text{ and } M \to N\gamma \text{ as } T/\theta_M \to 0 \tag{7.29}$$

Thus at $T \to 0$ all the spins line up in the direction of the field; the total magnetic moment is just N times the magnetic moment of one spin. This state is unique, so that the uncertainty and the entropy are both zero. We could attain the same effect at a fixed temperature by increasing the magnetic field, when again more spins would tend to align with the field. In the high temperature limit,

$$S \to Nk \ln 2 \text{ and } M \to 0 \text{ as } T/\theta_M \to \infty \tag{7.30}$$

Here we are in the state with maximum uncertainty, with all states being equally probable. The total number of arrangements for N spins in these circumstances is 2^N, which gives an uncertainty of $\ln(2^N) = N\ln 2$. Similarly if all states are equally probable, then up spins are not preferred to down spins so $N_\uparrow = N_\downarrow = N/2$ which means that $M = \gamma[\langle N_\uparrow\rangle - \langle N_\downarrow\rangle]$ and M must be zero.

7.5 Paramagnetic crystal in the grand canonical distribution

In this section we illustrate our claim that the choice of distribution is a matter of mathematical convenience only and that, apart from critical points, the result obtained from either of the distributions will be the same. We shall therefore calculate the properties of the model using the grand canonical distribution. We can do this rather easily since the expression for the grand partition function follows very easily from that for the canonical partition function. From equations (4.55), (4.65) and (7.22),

$$\Xi = \sum_{N=0}^{\infty} \left[\exp\left(\frac{\mu}{kT}\right) 2\cosh\left(\frac{\theta_M}{T}\right) \right]^N$$

which is a geometric series with a common ratio of

$$\exp\left(\frac{\mu}{kT}\right) 2\cosh\left(\frac{\theta_M}{T}\right)$$

which sums to

$$\Xi = \left[1 - 2\exp\left(\frac{\mu}{kT}\right) \cosh\left(\frac{\theta_M}{T}\right)\right]^{-1} \tag{7.31}$$

Now we use $Z = kT\ln\Xi$ and equation (7.12) to obtain

$$\mathcal{N} = \left(\frac{\partial Z}{\partial \mu}\right)_{B,T} = \frac{2\exp(\mu/kT)\cosh(\theta_M/T)}{1 - 2\exp(\mu/kT)\cosh(\theta_M/T)} \tag{7.32}$$

This can be solved for $\exp(\mu/kT)$ to give

$$\exp(\mu/kT) = \frac{\mathcal{N}}{2(\mathcal{N}+1)} \operatorname{sech}(\theta_M/T) \tag{7.33}$$

If we take the thermodynamic limit of $\mathcal{N} \to \infty$, we find

$$\exp(\mu/kT) = \tfrac{1}{2}(1 - 1/\mathcal{N})\operatorname{sech}(\theta_M/T) + O(1/\mathcal{N}^2) \tag{7.34}$$

From equations (7.13), (7.14) and (7.31) we get the other thermodynamic quantities

$$S = -k\frac{2\exp(\mu/kT)\cosh(\theta_M/kT)}{1 - 2\exp(\mu/kT)\cosh(\theta_M/T)}\left\{\left[\left(\frac{\mu}{kT}\right)\right.\right.$$
$$\left.\left. + \left(\frac{\theta_M}{T}\right)\tanh\left(\frac{\theta_M}{T}\right)\right] - k\ln\left[1 - 2\exp\left(\frac{\mu}{kT}\right)\cosh\left(\frac{\theta_M}{T}\right)\right]\right\}$$
(7.35)

and

$$M = \frac{2\gamma\exp(\mu/kT)\sinh(\theta_M/T)}{1 - 2\exp(\mu/kT)\cosh(\theta_M/T)} \tag{7.36}$$

If we now use equation (7.34) for $\exp(\mu/kT)$ these expressions reduce to

$$S = \mathcal{N}k\left\{\ln\left[2\cosh\left(\frac{\theta_M}{T}\right)\right] - \left(\frac{\theta_M}{T}\right)\tanh\left(\frac{\theta_M}{T}\right) + O(1/\mathcal{N}^2)\right\} \tag{7.37}$$

and

$$M = \mathcal{N}\gamma\tanh\left(\frac{\theta_M}{T}\right) \tag{7.38}$$

If we equate $\mathcal{N}$, the mean number of particles, with N, the given number of particles, (7.37) and (7.38) agree with the results for the canonical distribution to within $\sim 1/\mathcal{N}$. In fact that magnetization M agrees exactly once we equate $\mathcal{N}$ to N. We must remember however that $\mathcal{N}$ and N can differ by $\sim 1/\mathcal{N}^{1/2}$.

It should be noted that if the total Hamiltonian can be approximated to the sum of a vibrational Hamiltonian and a magnetic Hamiltonian without any terms linking the two, the magnetic contribution to the thermodynamic properties is just added to the other contribution. Thus for instance, the heat capacity of a paramagnetic Einstein solid in a field B would be the sum of the two contributions i.e.,

$$C(\text{total}) = C(\text{Einstein}) + C_B$$

The discussion in this chapter has been entirely concerned with non-interacting spins which are fixed in space. If we relax the second of these restrictions, then we lose the distinguishability of the spins. An example of such a system of indistinguishable non-interacting spins is a gas of low-density electrons. This case of electron paramagnetism is discussed in Chapter 12. The more difficult case is to relax the restriction of the spins being non-interacting. In the last twenty years there has been considerable progress in this topic and an introduction to that research is the subject of the last chapter.

110   MODELS OF MAGNETIC CRYSTALS

CHAPTER 7: SUMMARY OF MAIN POINTS

1. Magnetic solids are considered in the simplest way as arrays of atomic magnetic point dipoles.
2. The basic thermodynamic description is developed for use with both the canonical distribution and the grand canonical distribution. Attention is restricted to dipoles which for quantum-mechanical reasons have only two possible orientations with respect to the direction of an applied field.
3. Both distributions, canonical and grand canonical, are analyzed and it is shown that, for example, the predictions for entropy and the magnetization do not depend on the choice of distribution, at least to within terms of order $1/N$.

QUESTIONS

7.1 Obtain the equations linking the magnetic properties of a system with the number of accessible states in the microcanonical distribution.

7.2 Calculate the magnetic properties of the model discussed in this chapter using the microcanonical distribution. Start by showing that the number of states in the range $(U - \Delta U/2) < U < (U + \Delta U/2)$ is given approximately by

$$\Omega(U, \Delta U, B) = \frac{N!}{(N - N'_\uparrow)! N'_\uparrow!} \frac{\Delta U}{\gamma B}$$

where $N'_\uparrow$ is given by

$$U = -N'_\uparrow \gamma B + (N - N'_\uparrow)\gamma B$$

Hence using Stirling's approximation $\ln x! \simeq x \ln x - x$, for large x, show that

$$\ln \Omega \simeq N \ln N - \left(\frac{N}{2} + \frac{U}{2\gamma B}\right) \ln \left(\frac{U}{2\gamma B} + \frac{N}{2}\right)$$
$$- \left(\frac{N}{2} - \frac{U}{2\gamma B}\right) \ln \left(\frac{N}{2} - \frac{U}{2\gamma B}\right) - \ln \Delta U - \ln \gamma B$$

where for large N the last two terms can be neglected. Finally, using the results of Question 7.1 find M, S, etc.

7.3 For the model discussed in this chapter find the fractional deviation in energy in the canonical ensemble. Use this result to discuss the results of Question 7.2 and their relation to similar results in the canonical ensemble.

7.4 If the magnetic susceptibility of a solid is defined as $\chi = \mu_0 (\partial M/\partial B)_T$, use equation (7.26) to find χ and discuss the precise conditions for which χ becomes a constant independent of B.

7.5 A paramagnetic crystal (for example, cerium magnesium nitrate) contains N/V weakly interacting magnetic ions per unit volume. When placed in an applied magnetic field B, each ion has, according to quantum mechanics, a range of $(2J + 1)$ possible energies $-m_J\gamma B$ where m_J (the magnetic quantum number) runs from $-J$ to $+J$.

(a) Show that the canonical partition function $Q(B, T)$ is given by

$$Q(B,T) = \left[\frac{\sinh\left(\frac{(2J+1)y}{2J}\right)}{\sinh\left(\frac{y}{2J}\right)} \right]^N$$

where $y = J\gamma B/kT$.

(b) Show that when y is small the susceptibility reduces to

$$\chi = \mu_0 N \gamma^2 \frac{J(J+1)}{3kTV}$$

7.6 Prove that for a crystal containing N/V electric dipoles per unit volume in an *electric* field E, the polarization P is given by

$$P = \frac{NkT}{V} \left(\frac{\partial \ln Q}{\partial E} \right)_{T,N}$$

Show further that for sufficiently small E, P is proportional to E and find the next approximation. Note the similarity with the last question, but remember that quantum mechanics does not require the quantization of orientation of electric dipoles, since they are not in general associated with angular momentum. Thus, although the energy of each dipole p can only be in the range $-pE$ to $+pE$, a continuum of intermediate energies is allowed, so that sums should be replaced by appropriate integrals.

7.7 For the certain paramagnetic solid $\gamma = 9.273 \times 10^{-4}$ amperes m^{-2} and $s = \frac{1}{2}$. It is placed in a field of 6 teslas (6×10^4 gauss). To what temperature must the solid be cooled for 90 % of the atoms to be polarized in the direction of the field? The solid is removed from the field and replaced by an organic material which contains no paramagnetic atoms except hydrogen for which $s = \frac{1}{2}$ and $\gamma = 1.41 \times 10^{-26}$ amperes m^{-2}. To what temperature does this have to be cooled for 90 % of the hydrogen atoms to be polarized in the direction of the field?

7.8 Calculate the fractional deviation in M for the canonical ensemble. For a material containing 10^{18} magnetic ions mm^{-3}, at what volume will the deviation in M be 1 % of M, if $\gamma B/kT = 0.5$?

CHAPTER 8

Systems Composed of Non-interacting Sub-systems

8.1 Many-particle states
8.2 Distinguishable particles
8.3 Indistinguishable particles
    Question

8.1 Many-particle states

We have seen that the our main problem is to find the solutions of the Schrödinger equation and in particular the energy levels and the degeneracy of each level. Knowing these quantities we then have a straightforward procedure for calculating the macroscopic parameters. The trouble is that in only a few cases can we solve the Schrödinger equation exactly. Thus, in most cases, we have to find approximate expressions for the energy levels, although it is worthwhile emphasizing that this is an approximation in finding the properties of the accessible states and not in the statistical inference.

A class of systems where the Schrödinger equation can always be solved is one in which the systems are each composed of N non-interacting sub-systems. Examples of such systems are the Einstein or Debye solid, the sub-systems being the individual site-atoms or modes respectively, and the Maxwell–Boltzmann gas. We shall for brevity refer to the sub-systems as particles, although in some applications a more appropriate description might be 'quasi-particles'. For such systems the Hamiltonian may be written

$$H = \sum_{\alpha=1}^{N} h(\alpha) \tag{8.1}$$

where $h(\alpha)$ is the Hamiltonian for the single particle α. Thus, for a single particle

$$h(\alpha)\varphi_s(\alpha) = \epsilon_s \varphi_s(\alpha) \tag{8.2}$$

where φ_s are the single-particle eigenfunctions corresponding to the single-particle energy level ϵ_s. We have to solve the many-particle Schrödinger equation

$$H\psi = \sum_{\alpha=1}^{N} h(\alpha)\psi = E\psi \tag{8.3}$$

One can easily see by direct substitution that a solution is

$$\psi = \varphi_i(1)\varphi_j(2)\varphi_k(3)\ldots\varphi_z(N)$$

and

$$E = \epsilon_i(1) + \epsilon_j(2) + \epsilon_k(3) + \ldots \epsilon_z(N) \tag{8.4}$$

Here the subscripts $(i, j, k, \ldots s \ldots)$ refer to the single-particle states and the numbers $(1, 2, 3, \ldots \alpha \ldots N)$ label the particles. We can generate another solution by changing one or more of the single-particle states and provided the new states have different single-particle energies, i.e., we do not choose new single-particle states which are degenerate with the original states, the new many-particle state will have a different energy to the original one. If we have chosen degenerate single-particle states, then the many-particle state will be degenerate with the original. At first sight this seems to be the way to build up all the possible states of the many-particle system, but there are other possible states that cannot be generated by this procedure. In the state given by equation (8.4) interchanging particles 1 and 2 will apparently leave the energy unchanged but give us a different wave function. Whether this should be counted as a new state depends on whether we can distinguish between the particles. If we cannot distinguish them, then the labels 1 and 2 are just convenient, but physically there is no difference between these two states. On the other hand if the particles are distinguishable then we do indeed have two different degenerate states. To illustrate this important point, consider an example in which two particles (α and β) may occupy single-particle states with the energies ϵ_r and ϵ_s. The possible many-particle states are illustrated diagrammatically in Figure 8.1 and there are four of them.

However, if in Figure 8.1 α and β are indistinguishable, then there is no difference at all between the third and fourth states, and that configuration should be counted only *once*. Consequently, for indistinguishable particles there are, in this simple example, only three many-particle states

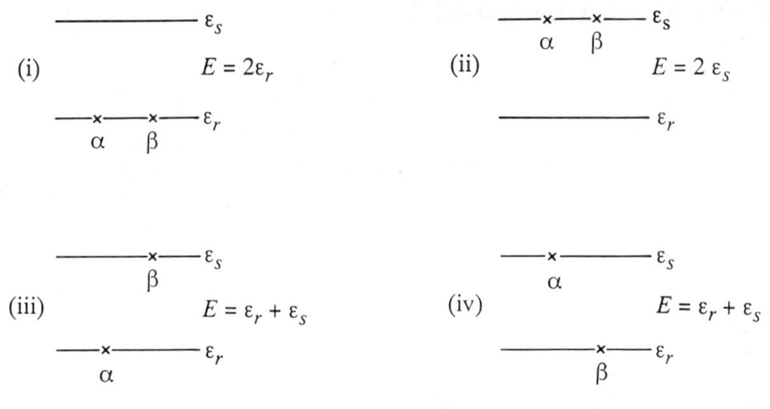

MANY-PARTICLE STATES

(i) $\quad\quad\quad\quad\quad\quad\quad$ —————— ϵ_s
$\quad\quad\quad\quad\quad\quad\quad\quad\quad\quad\quad\quad\quad E = 2\epsilon_r$
$\quad$ —×———×—— ϵ_r
$\quad\quad\alpha\quad\quad\beta$

(ii) $\quad$ —×———×—— ϵ_s
$\quad\quad\quad\alpha\quad\quad\beta$
$\quad\quad\quad\quad\quad\quad\quad\quad E = 2\epsilon_s$
$\quad\quad\quad\quad\quad\quad$ —————— ϵ_r

(iii) $\quad$ —×—— ϵ_s
$\quad\quad\quad\beta$
$\quad\quad\quad\quad\quad\quad\quad E = \epsilon_r + \epsilon_s$
$\quad$ —×—— ϵ_r
$\quad\quad\alpha$

(iv) $\quad$ —×—— ϵ_s
$\quad\quad\alpha$
$\quad\quad\quad\quad\quad\quad\quad E = \epsilon_r + \epsilon_s$
$\quad$ —×—— ϵ_r
$\quad\quad\beta$

Figure 8.1 The diagram represents schematically the possible states of two particles α and β distributed between two single-particle levels ϵ_r and ϵ_s. The single-particle levels are assumed to be non-degenerate. If the particles are indistinguishable then states (iii) and (iv) are identical and should only be counted once. If in addition the particles obey Fermi–Dirac statistics, states (i) and (ii) are excluded. Thus, for distinguishable particles there are four states, for bosons three states and for fermions one state.

instead of four. Thus, rendering particles indistinguishable has the effect of reducing the degeneracy associated with the energy $(\epsilon_r + \epsilon_s)$ from two to one. This reduction may be very considerable where many particles and single-particle states are involved as we shall see. A further effect arises for indistinguishable particles (for example, electrons) which are also subject to the Pauli exclusion principle, since then no multiple-occupation is permitted and, in our simple example only *one* of the original four two-particle states remains, with energy $(\epsilon_r + \epsilon_s)$.

We will discuss these effects in greater detail in subsequent sections, but we note that whether the particles are distinguishable or not, the states can be specified by a set of integers known as occupation numbers. These integers specify how many particles are in, or occupy each of the single-particle states. Consider the above example, for distinguishable particles the four states are specified as $[n_r = 2; n_s = 0]$, $[n_r = 1; n_s = 1]$ (2-fold degenerate), and $[n_r = 0; n_s = 2]$. For particles obeying the Pauli exclusion principle the only allowed state is $[n_r = 1; n_s = 1]$ with no degeneracy. In general, the energy of the system will be given by

$$E_{\{n_s\}} = \sum_s n_s \epsilon_s \tag{8.5}$$

If we are considering a fixed number of particles then we must have $\Sigma_s n_s = N$. For the rest of this chapter we discuss the two cases, distinguishable and indistinguishable particles, separately.

8.2 Distinguishable particles

For a system of N distinguishable particles the energy levels are specified by the set of occupation numbers n_s such that $\Sigma_s n_s = N$ and the n_s can take the values $n_s = 0, 1, 2, \ldots, N$. This set of integers specifies the energy level uniquely but not the quantum-mechanical state. Since the particles are distinguishable, the number of states having the same energy, i.e., the degeneracy, is the number of ways we can choose the set of integers $\{n_s\}$ from the number N, such that $\Sigma_s n_s = N$. This number is easily found (see Question 8.1), and we have for the degeneracy of the energy level $E_{\{n_s\}}$

$$g_{\{n_s\}} = \frac{N!}{\prod_s n_s!}, \quad \text{where} \quad \sum n_s = N \tag{8.6}$$

The canonical partition function may be immediately written down:

$$Q(N) = \sum_{\{n_s\}}{}' g_{\{n_s\}} \exp\left(-\frac{E_{\{n_s\}}}{kT}\right) \tag{8.7}$$

$$= \sum_{\{n_s\}}{}' \frac{N!}{\prod_s n_s!} \exp\left(-\sum_s \frac{n_s \epsilon_s}{kT}\right)$$

$$= \sum_{\{n_s\}}{}' \frac{N!}{\prod_s n_s!} \prod_s \left[\exp\left(-\frac{\epsilon_s}{kT}\right)\right]^{n_s} \tag{8.8}$$

where the primed sum Σ' indicates that all n_s can run from 0 to N subject to the constraint $\Sigma_s n_s = N$. Now equation (8.8) is greatly simplified by use of the multinomial theorem which can be written in the form:

$$(x_1 + x_2 + \ldots + x_s + \ldots)^N \equiv \sum_{\{n_s\}}{}' \frac{N!}{\prod_s n_s!} x_1^{n_1} x_2^{n_2} x_3^{n_3} \ldots x_s^{n_s} \tag{8.9}$$

Taking $x_s \equiv \exp(-\epsilon_s/kT)$, the right-hand side of this becomes identical with equation (8.8) which can, therefore, be written

$$Q(N) = \left[\sum_s \exp\left(-\frac{\epsilon_s}{kT}\right)\right]^N \tag{8.10}$$

This result might have been expected on the grounds of the simple argument given in Section 5.6. Conversely, the rigorous algebra presented above justifies that section (see also Appendix 3). The bridge equation for the canonical distribution gives

$$F = -kT \ln Q = -NkT \ln \left[\sum_s \exp\left(-\frac{\epsilon_s}{kT}\right) \right] \quad (8.11)$$

Equation (8.10) says that for N non-interacting distinguishable particles, the partition function is the partition function for one particle raised to the power N. As a consequence F is proportional to N, as it should be. All other thermodynamic quantities can be derived from equation (8.11); in particular

$$S = -\left(\frac{\partial F}{\partial T}\right)_V$$

$$= \frac{N}{T} \frac{\sum_s \epsilon_s \exp\left(-\frac{\epsilon_s}{kT}\right)}{\sum_s \exp\left(-\frac{\epsilon_s}{kT}\right)} + Nk \ln \sum_s \exp\left(-\frac{\epsilon_s}{kT}\right) \quad (8.12)$$

After a little manipulation this can be written as

$$S = -Nk \sum_s p_s \ln p_s \quad (8.13)$$

where

$$p_s \equiv \frac{\exp\left(-\frac{\epsilon_s}{kT}\right)}{\sum_s \exp\left(-\frac{\epsilon_s}{kT}\right)} \quad (8.14)$$

is the probability of a particle being in the single-particle state s. Since the molecules are all independent the probability of a particular many-particle state is the product of the probabilities for the single-particle states comprising the many-particle state. Under these conditions, the uncertainty H, of the many-particle state is the sum of the uncertainties for the single-particle states (see Section 3.2). Since the uncertainty in the state of a single particle is $-\Sigma_s p_s \ln p_s$ and $S = kH$, equation (8.13) is the result we would expect.

Although the probability p_s of a particle being in a state s is a useful quantity to introduce, it is important to realize that it has none of the properties we have associated with all other probability functions, namely that it should be sharply peaked and that the observed value should be identical with the mean value. This is simply because if we are to measure the individual energies of a few particles we would not expect to get the same energy for each one. If one considers the fractional deviation in the energy of a single particle, one finds a result like equation (4.41):

$$\langle \Delta \epsilon \rangle^2_{\text{rms}} = \sum_s p_s \epsilon_s^2 - \left(\sum_s p_s \epsilon_s \right)^2 = kT^2 c_V \tag{8.15}$$

where c_V is the heat capacity per particle. Hence the fractional deviation is

$$\frac{\langle \Delta \epsilon \rangle_{\text{rms}}}{\langle \epsilon \rangle} = \frac{kT}{u} \left(\frac{c_V}{k} \right)^{1/2} \tag{8.16}$$

where u is the mean energy per particle. Typically, $c_V \sim k$ and $u \sim kT$ so that

$$\frac{\langle \Delta \epsilon \rangle_{\text{rms}}}{\langle \epsilon \rangle} \sim 1 \tag{8.17}$$

Hence fluctuations in energy around u are not typically small. These remarks parallel earlier discussion in Chapter 5, where the fractional deviations were calculated for several different single-particle sub-systems.

8.3 Indistinguishable particles

The energy levels of a system of identical particles have the same form as that for distinguishable particles

$$E_{\{n_s\}} = \sum_s n_s \epsilon_s \tag{8.18}$$

However, for the case of indistinguishable particles, there is no degeneracy due to rearrangements among particles, since these are physically meaningless. Thus, not only are the energy levels specified by the set of occupation numbers $\{n_s\}$, but they also uniquely determine the state. Moreover, since the particles are indistinguishable, the Hamiltonian is symmetric with respect to the interchange of any two particles, that is

$$H(r_1, r_2 \ldots r_\alpha, r_\beta \ldots r_N) = H(r_1, r_2 \ldots r_\beta, r_\alpha \ldots r_N) \tag{8.19}$$

As a consequence of this it is easy to show that the solutions of the Schrödinger equation are either symmetric or anti-symmetric with respect to the interchange of two particles, that is,

$$\psi(r_1, r_2 \ldots r_\alpha, r_\beta \ldots r_N) = \pm \psi(r_1, r_2 \ldots r_\beta, r_\alpha \ldots r_N) \quad (8.20)$$

For both of these solutions, $|\psi|^2$, which is the physically significant quantity, is invariant to the permutation of the particles, as it should be if the particles are identical. In nature it is found that there are two types of particles. The first type, called fermions, are such that the wavefunction for two or more particles is anti-symmetric under the interchange of two particles, i.e., the minus sign is appropriate in equation (8.20). An immediate consequence of this is that the occupation numbers n_s can take only the values 0 and 1. The proof of this statement is beyond the scope of this book, but we can demonstrate its reasonableness by considering the wave function for two non-interacting fermions. An anti-symmetric function would be

$$\psi(r_1, r_2) = \frac{+1}{\sqrt{2}} [\varphi_1(r_1)\varphi_2(r_2) - \varphi_1(r_2)\varphi_2(r_1)]$$

$$= \frac{-1}{\sqrt{2}} [\varphi_1(r_2)\varphi_2(r_1) - \varphi_1(r_1)\varphi_2(r_2)] \quad (8.21)$$

In the occupation number representation this corresponds to $n_1 = n_2 = 1$, all other n_s being zero. However, if we try to construct a wavefunction for which $n_1 = 2$ and all other n_s are zero, we obtain

$$\psi(r_1, r_2) = \frac{1}{\sqrt{2}} [\varphi_1(r_1)\varphi_1(r_2) - \varphi_1(r_2)\varphi_1(r_1)] = 0 \quad (8.22)$$

The physical meaning of this zero is that double occupation of a single-particle state is forbidden or excluded; this is a simple example of the Pauli exclusion principle.

The second type of particles are called bosons, whose wavefunctions for two or more particles are symmetric under the permutation of particles, i.e., the plus sign is appropriate in equation (8.20). For bosons there is no limit on the value of the n_s. To summarize, the quantum-mechanical state of a system of identical particles can be specified uniquely by a set of occupation numbers. For fermions, whose wave function is antisymmetric under the permutation of particles, these occupation numbers are limited to the values 0 and 1. For bosons whose wave function is symmetric under permutations, n_s can take any value between 0 and N (the total number of particles).

We are now in a position to write down the canonical partition function:

$$Q(N) = \sum_{\{n_s\}}^{X} {}' \exp\left(-\sum_s \frac{n_s \epsilon_s}{kT}\right) \quad (8.23)$$

$$= \sum_{\{n_s\}}^{X}{}' \prod_s \left[\exp\left(-\frac{\epsilon_s}{kT}\right) \right]^{n_s} \tag{8.24}$$

where, as before, the primed sum denotes summation only over sets $\{n_s\}$ which are constrained by the condition $\Sigma_s n_s = N$. The symbol X denotes upper limiting value allowed to any of the n_s, $X = 1$ for fermions, $X = N$ for bosons. The degeneracy factor which appeared in the corresponding expression (8.8) for distinguishable particles, is absent in (8.24). As a consequence, the resulting sums cannot be performed explicitly. This difficulty arises from the restriction $\Sigma_s n_s = N$. We shall see in Chapter 10 that the removal of this restriction is possible by using the grand canonical distribution, which makes possible an explicit summation. Meanwhile, under physical conditions which are quite commonly met in gases, an excellent approximation to the summation in equation (8.24) can be made. This case, the perfect Maxwell–Boltzmann gas, is the subject of Chapter 9.

CHAPTER 8: SUMMARY OF MAIN POINTS

1. For a system of N non-interacting particles, the energy level of the system is the sum of the energies of the individual particles.
2. The state of the system is specified by the set of occupation numbers, that is, the number of particles in a single-particle state.
3. The number of system states is determined by whether the particles are distinguishable or are non-distinguishable.
4. If the particles are indistinguishable, then they are either classified as obeying Bose statistics or Fermi statistics. For particles obeying Bose statistics the occupation numbers are unrestricted beyond the conservation requirement that $\Sigma_s n_s = N$, but for particles obeying Fermi statistics the occupation numbers are restricted to the values 0 to 1.
5. For distinguishable particles, it is possible to evaluate the canonical partition function exactly.

QUESTION

8.1 Suppose we have N distinguishable balls and a number of boxes. The first box is such that we cannot put any balls in it, the second we can put one ball, the third two. In general, in the sth box we can put n_s balls. Suppose with our N balls we decide to fill a given r of the boxes in such a way that $\Sigma_s n_s = N$. The question is this: in how many ways can we do it? Show that, since we can choose the first ball in N ways, the second in $(N-1)$ ways, etc., the total number of ways of choosing the balls is $N!$. Show also that the total number of rearrangements within the sth box is $n_s!$. Hence deduce that the total number of ways of filling the r boxes is

$$\frac{N!}{\prod_s n_s!} \quad \text{where} \quad \sum_s n_s = N$$

CHAPTER 9
Aspects of the Classical Gas

9.1 The high-temperature low-density criterion
9.2 The thermodynamic properties of a monatomic gas
9.3 Thermodynamic properties of polyatomic gases
9.4 Non-reacting gas mixtures
9.5 Reacting gas mixtures and dissociation
9.6 Maxwell–Boltzmann energy distribution
Questions
References

9.1 The high-temperature low-density criterion

We have shown in Chapter 8 that the canonical partition function for a system composed of non-interacting sub-systems takes different forms, depending on whether the sub-systems are distinguishable or indistinguishable in the strict quantum-mechanical sense. Thus, for distinguishable sub-systems, equation (8.8) factorizes as shown by equation (8.10). Thus

$$Q_{\text{dist}}(N) = \sum_{\{n_s\}}{}' \frac{N!}{\prod_s n_s} \prod_s \left[\exp\left(-\frac{\epsilon_s}{kT}\right) \right]^{n_s} = q^N \qquad (9.1)$$

whereas, from equation (8.24), for indistinguishable sub-systems

$$Q_{\text{indist}}(N) = \sum_{\{n_s\}}{}^X \prod_s \left[\exp\left(-\frac{\epsilon_s}{kT}\right) \right]^{n_s} \qquad (9.2)$$

THE HIGH-TEMPERATURE LOW-DENSITY CRITERION

The upper limit X in (9.2) takes the appropriate value, 1 for fermions or N for bosons.

Equation (9.2) does not in general factorize, as the reader can confirm in a simple example by considering the contributions to the partition function from the number of possible microstates associated with the distributions of two particles among two states as illustrated in Figure 8.1. If the particles are distinguishable, all four microstates contribute a term in the sum for Q_{dist} and the result readily factorizes. If the particles are indistinguishable bosons, only three of the four are valid and Q_{indist} is then a three-term sum which does not factorize. If the particles are indistinguishable fermions, the calculation is again different and Q_{indist} has only one term corresponding to a particle in each of the two available single-particle states.

We now consider a slightly more elaborate example, that of three distinguishable particles, each with access to 1000 single-particle states. There are $(1000)^3 = 10^9$ microstates in all, of which a sizeable number $1000 \times 999 \times 998$, i.e., 99.7 %, are free of multiple occupation. Now we imagine 'switching on' the indistinguishability of the particles (this is of course not possible in reality but that does not invalidate the logic of the argument). Those microstates which do not have multiple occupation (i.e., the great majority) are easily dealt with in that their contribution to Q has to be reduced by a factor 3! because rearrangements among the three particles no longer generate distinct states. Now the factorization $Q = q^N$ depends on *all* microstates being included in the sum, but some have to be excluded because of indistinguishability and (for fermions only) because of the exclusion principle. The recipe suggested above, that is, simply dividing by $N!$, is clearly only valid for those microstates which, if we had distinguishable particles, would be free of multiple occupation. However, if we can identify a physical regime in which the probability of multiple occupation of single-particle states is vanishingly small (i.e., the probability $P(n_s)$ of occupation of a state s by n_s particles is significant for $n_s = 0$ or $n_s = 1$ but vanishingly small for higher values), then $\Pi_s n_s! \approx 1$ and $Q_{\text{indist}}(N) = Q_{\text{dist}}(N)/N!$. Furthermore, in such a regime the distinction between fermion and boson would be unimportant, since n_s would rarely exceed 1 anyway. Is such a regime physically obtainable? If so, the system must be such that there are far more single-particle *states* than particles with energies in any range $\Delta \epsilon_s$, so that the average occupation number is well below unity. We now obtain expressions for both these quantities for the case of a gas.

It was shown in Section 5.5 that the single-particle energies for a particle of mass m in a volume V are given by

$$\epsilon = \frac{h^2}{8mV^{2/3}} (n_x^2 + n_y^2 + n_z^2) \tag{9.3}$$

$$= k\theta_t (n_x^2 + n_y^2 + n_z^2) \tag{9.4}$$

where n_x, n_y, n_z are non-zero positive integers and $\theta_t \equiv h^2/8mkV^{2/3}$ is a convenient characteristic temperature, which is typically so tiny as to be well below any physically accessible temperature. For example, for a helium atom in a volume of one litre, $\theta_t = 5.95 \times 10^{-17} K$! Taking note of the fact that equation (9.4) is the equation to the positive octant of a spherical surface of radius $R = (\epsilon/k\theta_t)^{1/2}$ in n_x, n_y, n_z space we can say that states with energy ϵ are represented by points on the curved surface of that octant. Moreover, the representation of *all* accessible states will produce a simple cubic lattice of points with unit spacing in n_x, n_y, n_z-space. Consequently, a state corresponds to unit volume in that space, so that the number ρ of states with energies in the range 0 to ϵ is approximately equal to the volume of the octant, i.e. $\frac{1}{6}\pi(\epsilon/k\theta_t)^{3/2}$. Clearly, the approximation is better for higher values of $\epsilon/k\theta_t$, since the discreteness of the unit volume being counted becomes less important when there are many of them. For example, for $\epsilon/k\theta_t = 10^4$ an exact count yields 511776 states which compares quite closely with the approximate formula which gives 523599. Since it has been noted above that θ_t can easily be as small as 10^{-17} K for macroscopic gas samples, the approximation can be expected to be excellent. Thus

$$\rho = \frac{\pi}{6}\left(\frac{\epsilon}{k\theta_t}\right)^{3/2} = \frac{4\pi Vg}{3h^3}(2m\epsilon)^{3/2} \tag{9.5}$$

where ρ is the number of states with energies between 0 and ϵ, and g is included as a possible spin degeneracy (i.e., $g \equiv 2S + 1$ where S is the total spin quantum number of the particles). Now, treating ϵ as a quasi-continuous variable (which is clearly valid since a choice of ϵ can readily be made which satisfies $d\epsilon/\epsilon \ll 1$ but also encompasses a large number of states because of the smallness of θ_t), we differentiate equation (9.5) to obtain

$$d\rho = \frac{2\pi Vg}{h^3}(2m)^{3/2}\epsilon^{1/2}d\epsilon \tag{9.6}$$

where $d\rho$ is the number of states having energies in the range $d\epsilon$. Next we wish to obtain, for comparison with $d\rho$, the number dn of particles having energies in the range $d\epsilon$. This may be taken as an experimentally determined quantity which is known to agree for a wide variety of gases and conditions with the formula

$$dn = \frac{2\pi N}{(\pi kT)^{3/2}}\epsilon^{1/2}\exp\left(\frac{-\epsilon}{kT}\right)d\epsilon \tag{9.7}$$

This is, of course, the Maxwell–Boltzmann distribution which we shall later derive *from* statistical mechanics. However, at this point, since we aim only

THE HIGH-TEMPERATURE LOW-DENSITY CRITERION   125

at a rough-and-ready criterion of rare multiple occupation, we take equation (9.7) as an empirical description of experimental results.[1] Thus, the required criterion for a given dϵ is

$$dn \ll d\rho \tag{9.8}$$

Using equations (9.6) and (9.7) it is possible to write this condition as

$$\frac{Nh^3}{Vg(2\pi mkT)^{3/2}} \exp\left(\frac{-\epsilon}{kT}\right) \ll 1 \tag{9.9}$$

Now we wish this criterion to be true for all possible energies ϵ, and we observe that if it is true for the ground-state energy ($3k\theta_t$) then it is bound to be true for all other (i.e., higher) energies. But we know that θ_t is typically far below any attainable temperature by a very large factor, so that in (9.9) the exponential term for the ground-state energy is always close to unity. Consequently, equation (9.9) can safely be rearranged in the form of a temperature criterion:

$$T \gg \theta_0 \equiv \frac{h^2}{2\pi mk}\left(\frac{N}{gV}\right)^{2/3} = \frac{4}{\pi}\left(\frac{N}{g}\right)^{2/3} \theta_t \tag{9.10}$$

The meaning of this inequality is that it defines a physical regime (in terms of a characteristic temperature θ_0) in which n_s is extremely unlikely to take values other than 0 or 1. This is the regime we are seeking, in which no single-particle state is likely to be occupied by more than one particle. In these circumstances $\Pi_s n_s! \approx 1$ and

$$Q_{\text{indist}}(N) = \frac{1}{N!} Q_{\text{dist}}(N) \tag{9.11}$$

$$= \frac{1}{N!} q^N \tag{9.12}$$

where, from the arguments given in Section 5.5,

$$q \equiv \sum_s \exp\left(-\frac{\epsilon_s}{kT}\right) = \left(\frac{2\pi mkT}{h^2}\right)^{3/2} gV \tag{9.13}$$

Also, using Stirling's approximation, we may write

$$\frac{1}{N!} \approx \left(\frac{e}{N}\right)^N \tag{9.14}$$

Thus

$$Q_{\text{indist}}(N) = \left[\frac{2\pi mkT}{h^2}\right]^{3N/2} \left(\frac{egV}{N}\right)^N \tag{9.15}$$

provided that $T \gg \theta_0$ where θ_0 is defined in equation (9.10). There are two interesting points to note about the characteristic temperature θ_0. First, since it is equal to $\theta_t N^{2/3}$ (apart from a numerical factor close to unity), it is much greater than θ_t for macroscopic systems. It follows that in evaluating the single-particle partition function, the replacement of the sum by an integral is amply justified. The second point is that we may interpret equation (9.10) physically as the requirement that if the particles are to be distinct entities, the wave packets associated with them should not overlap. We can see this by using the uncertainty relation to estimate the size of the wave packets:

$$\Delta x \sim \frac{\hbar}{\Delta p} \tag{9.16}$$

where

$$\Delta p = [\langle p^2 \rangle - \langle p \rangle^2]^{1/2} \tag{9.17}$$
$$= (\langle p^2 \rangle)^{1/2} \tag{9.18}$$

The last step follows from the fact that $\langle p \rangle = 0$. Moreover, taking $p^2 = 2m\epsilon$ for each particle and using equation (9.7) we obtain

$$\Delta p = (3mkT)^{1/2} \tag{9.19}$$

Consequently, the approximate volume associated with each moving particle is

$$v = (\Delta x)^3 \approx \frac{\hbar^3}{(3mkT)^{3/2}} \tag{9.20}$$

This should now be required to be much less than the free volume available to each particle, namely V/N. The criterion

$$v \ll \frac{V}{N} \tag{9.21}$$

can be arranged to read

$$T \gg \frac{\hbar^2}{3mk}\left(\frac{N}{V}\right)^{2/3} \tag{9.22}$$

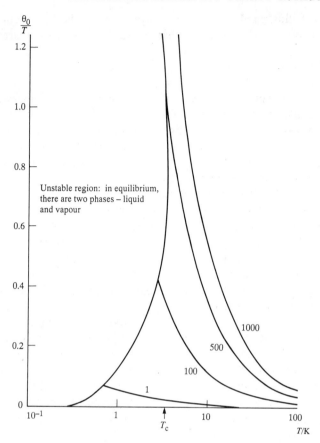

Figure 9.1 A plot of θ_0/T versus T for gaseous ^{3}He. The left-hand curve describes condensation ($T_c = 3.4$ K). The various labelled right-hand curves show the behaviour for different number densities ($1 \equiv 2.69 \times 10^{25}$ atoms m^{-3}, the NTP value). It can be seen that significant breaches of criterion (9.22) can only be achieved with number densities in excess of about 10^{28} atoms m^{-3}, when it would in any case no longer be proper to ignore interaction between atoms.

This condition is almost exactly the same as that given in equation (9.10), but has been obtained by a different and possibly more physical argument.

For macroscopic samples of real gases it is impossible to break condition (9.10) without applying such pressures that the whole model becomes invalid through neglect of the potential energy of interaction. The most likely gas with which to attempt to breach condition (9.10) would seem to be ^{3}He because of the combination of low atomic mass and low critical temperature (making possible high number densities without condensation).

Our point is illustrated in Figure 9.1, which shows the behaviour of θ_0/T for various numbers of densities characterized by initial pressures (1 atm, 100 atm, 500 atm) at room temperature.

It is clear from Figure 9.1 that where liquid and vapour phases coexist, the number density in the vapour is such that $T \geqslant \theta_0$ throughout the temperature range $T \leqslant T_c$. In the single-phase region it may appear that, if the initial pressure at room temperature (before cooling) is greater than about 500 atmospheres, it might just be possible to breach condition (9.10). However, it must be remembered that the interatomic separation $\sim (V/N)^{1/3}$ is then only about 0.43 nm, which is so close that neglect of the potential energy of interaction is barely justifiable. We conclude that in describing gaseous ^{3}He we are always in the regime $\theta_0 < T$ for number densities which justify neglecting the interactions.

The only atoms existing whose masses are less than or equal to that of ^{3}He are ^{1}H (ordinary hydrogen), ^{2}H (or D for deuterium), and ^{3}H (or T for tritium). A gas composed of *atoms* of these isotopes is unstable even at quite high temperatures (see, for example, Section 9.5 below) and the atoms tend quickly to relax to form diatomic molecules. The interaction between the molecules is such that the gases condense at temperatures higher than those of the helium isotopes; the normal boiling points of ^{3}He, ^{4}He, and ^{1}H$_2$ are, respectively, 3.19 K, 4.21 K, and 20.3 K. This is why Figure 9.1 is plotted for ^{3}He rather than ^{1}H. In recent years a considerable research effort has been directed towards spin-polarized monatomic hydrogen which has such weak interatomic interaction that it may remain in gaseous form to very low temperatures, possibly even to the absolute zero. The experimental difficulties, and they are considerable, in keeping the gas in this form for convenient periods of time, involve the application of a strong magnetic field and close attention to the preparation of container surfaces where depolarization and the formation of molecules can otherwise easily occur. It is to be expected, although experimental verification is difficult to obtain, that such a gas may breach the criterion (9.22) if T is sufficiently low and N/V sufficiently high. In such a state, quantum degeneracy would be expected, with ^{1}H and ^{3}H atoms behaving as bosons and ^{2}H atoms behaving as fermions.[2]

9.2 The thermodynamic properties of a monatomic gas

Using the bridge equation for the canonical distribution, and equation (9.15), we get

$$F = -NkT \ln Q_{\text{indist}} = \tfrac{3}{2} NkT \ln \left(\frac{2\pi mkT}{h^2} \right) + NkT \ln \left(\frac{egV}{N} \right) \quad (9.23)$$

Hence

$$S = -\left(\frac{\partial F}{\partial T}\right)_V = \tfrac{3}{2}Nk + Nk\ln\left[\left(\frac{2\pi mkT}{h^2}\right)^{3/2}\frac{egV}{N}\right] \quad (9.24)$$

$$C_V = T\left(\frac{\partial S}{\partial T}\right)_V = \tfrac{3}{2}Nk \quad (9.25)$$

$$U = F + TS = \tfrac{3}{2}NkT \quad (9.26)$$

and

$$p = -\left(\frac{\partial F}{\partial V}\right)_T = \frac{NkT}{V} \quad (9.27)$$

These results, with the possible exception of (9.24), will be familiar from elementary kinetic theory. It should be noted that they are valid only in the temperature range $T \gg \theta_0$, so it is not legitimate to consider the limiting cases as T approaches absolute zero. This is the reason why the fact that expression (9.24) for the entropy appears inconsistent with the third law (in that it approaches $-\infty$) is not a cause for concern.

9.3 Thermodynamic properties of polyatomic gases

Molecules may have additional degrees of freedom, for example, vibrational and rotational. They may also have electronic transitions which will contribute to the thermodynamic functions. Thus we have to modify the partition function (9.15) and write

$$Q = \left(\frac{2\pi mkT}{h^2}\right)^{3N/2}\left(\frac{egV}{N}\right)^N q_r^N q_v^N q_e^N \quad (9.28)$$

where q_r, q_v, and q_e are the partition functions for the rotations, vibrations and electronic transitions of a single molecule, examples of which were given in Chapter 5. The fact that the partition function for the whole gas can be written in the form given by (9.28) was discussed in Section 5.6. Hence

$$F = -NkT\ln\left[\left(\frac{2\pi mkT}{h^2}\right)^{3/2}\left(\frac{egV}{N}\right)\right] + NkT\ln q_r$$
$$+ NkT\ln q_v + NkT\ln q_e \quad (9.29)$$

The contributions to the thermodynamic properties of the internal modes simply add to the monatomic contribution, with the appropriate mass inserted. The latter contribution is also known as the translational

130   ASPECTS OF THE CLASSICAL GAS

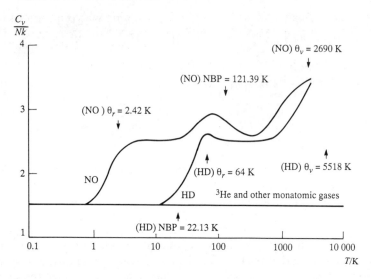

Figure 9.2 Heat capacities of two diatomic gases (NO and HD) compared with monatomic ^{3}He. The curves are theoretical, based on information on energy levels obtained from molecular spectroscopy and closely given by expressions derived in Section 9.3(a) and Section 9.3(b). The only difference is that at high temperatures, stretching and anharmonicity become significant. For thorough treatments see the original papers whose references are given in Section 9.3(a) and Section 9.3(b) and from which the curves were obtained. Direct experimental data, though still imperfect and incomplete, tends to confirm the predicted behaviour.

contribution. None of the other contributions depends on the volume, so the equation of state remains the same as for a monatomic gas.

A polyatomic molecule will in general have three rotational modes, each with a different moment of inertia; it may also have a number of vibrational modes each with a different frequency. We shall here restrict the discussion to diatomic heteronuclear molecules so that immediate use may be made of expressions derived in Chapter 5. Most of the discussion also applies to all diatomic homonuclear molecules such as N_2 with the important exception of low-temperature rotational contributions which are complicated by matters of symmetry. These arise because if the atoms in a homonuclear molecule are indistinguishable in *every* respect (including, importantly for hydrogen, orientation of nuclear spin), the rotation through π radians brings the molecule back to a configuration which is indistinguishable from its initial configuration. On the other hand where the atoms *are* distinguishable, only a rotation of 2π will be sufficient. This affects the summations in q_r. When it comes to comparison with experimental results at accessible temperatures, these matters affect only hydrogen H_2, and we can avoid the problem by referring to deuterium hydride (HD or $^1H^2H$) in which one atom in the molecule is deuterium (heavy hydrogen). Deuterium hydride shows very well the effects of rotation and vibration on heat capacity, but it has no electronic transitions with low enough energy to affect the heat capacity at accessible temperatures. For this reason we shall refer also to nitric oxide (NO) whose heat capacity, together with that of deuterium hydride, is shown in Figure 9.2.

(a) Deuterium hydride (HD)

This gas has $\theta_r = 64$ K, $\theta_v = 5518$ K, and a normal boiling point of 22.1 K, so it is experimentally possible to make measurements in the temperature range $T < \theta_r$ and indeed Clusius and Bartholomé[3] were able to go as low as 35 K. The upper limit is set by the dissociation of the molecules (see Section 9.5). At 1 atmosphere and 3000 K the gas is about 6% dissociated, so this temperature may be taken as the upper limit of validity of (9.29) for this gas. Thus HD will have translational, rotational, and vibrational contributions. The partition function (9.28) does not include interaction terms between the various degrees of freedom. The curve shown in Figure 9.2 was calculated by Johnston and Long[4] using a partition function which included terms in the Hamiltonian representing both the stretching of molecules in rotation and the anharmonic contributions to the vibrations. These extra terms are of little importance up to about 1000 K but account for the fact that at the highest temperatures, C_V apparently exceeds $7kT/2$. The small peak at $T \sim 50$ K arises from the rotational degree of freedom discussed in Section 5.4 (see also Question 5.6). From about 100 K to about 1000 K, the specific heat remains constant at $5Nk/2$. This is because θ_v is high, which means that at temperatures $T \lesssim 1000$ K the molecules behave as though they had no vibrational degrees of freedom and only two rotational degrees (because the molecule is collinear) and the usual three translational degrees.

(b) Nitric oxide (NO)

This gas has $\theta_r = 2.42$ K, $\theta_v = 2690$ K and a normal boiling point of 121.4 K. There is also an electronic transition for which $\Delta\epsilon/k = 178$ K which is the main reason for treating it here. The canonical partition function has an extra factor of q_e^N where q_e is given by equation (5.4). Its effect on the heat capacity is to add a term equal to N times equation (5.11) which has a maximum value of $0.439 Nk$ when $T = 0.417 \Delta\epsilon/k$. Figure 9.2 shows the calculated heat capacity[5] including stretching and anharmonic contributions. The rotational step is unobservable in NO as in all gases except hydrogen, so that the high-temperature limiting form, q_r, is always adequate for comparison with experiment. As for the electronic transition peak at 74 K, only its foothills have been observed[6] down to 128 K.

9.4 Non-reacting gas mixtures

Consider a mixture of gases, for each of which, separately, the approach in this chapter is valid. The total partition function is the product $Q_A Q_B Q_C \ldots$ if each of the gases is assumed independent. Thus, for example, for a mixture of two gases,

$$Q = \left(\frac{eq_A}{N_A}\right)^{N_A} \left(\frac{eq_B}{N_B}\right)^{N_B} \tag{9.30}$$

where q_A, q_B are single-molecule or single-atom partition functions, and N_A, N_B are the numbers present in V at T. Thus, from the bridge equation (4.37),

$$F = -kT \ln Q = -kT \left[N_A \ln \left(\frac{eq_A}{N_A}\right) + N_B \ln \left(\frac{eq_B}{N_B}\right) \right] \tag{9.31}$$

It readily follows (Question 9.2) that, for the mixture,

$$C_V = \frac{3}{2}(N_A + N_B)k \tag{9.32}$$

and

$$pV = (N_A + N_B)kT \tag{9.33}$$

which is well known as Dalton's law of partial pressures. Also S is the simple sum of the entropies of each gas separately in V at T. However, an interesting situation arises as gases are mixed by removal of a separating partition, a process which is readily shown to be accompanied by an *increase* in entropy. There is a logical puzzle here since this entropy does not apparently vanish when the gases being mixed are identical! This is the famous Gibbs' paradox, whose resolution is the topic of Question 9.3.

9.5 Reacting gas mixtures and dissociation

If the mixed gases are non-reactive, then little more of interest can be added to the ideal gas mixture discussed above. However, if A and B react together in some way, we may enquire into the nature of the equilibrium finally attained between A, B, and their reaction product(s). It was argued in Section 4.3 that at constant V and T, the free energy F is a minimum at equilibrium. This result can also be obtained by purely thermodynamic arguments (see Section 12.1). As an example of the power of this approach,

REACTING GAS MIXTURES AND DISSOCIATION   133

consider initially a gas of N_{HD} molecules of HD in a volume V at temperature T. There are several possibilities open to each molecule: it may continue as HD, dissociate into atomic H and D, recombine with other molecules to form H_2 and D_2, ionize by losing an electron, etc. These, and many other variations are all *possible* but we assert that they take place only to the extent that they reduce the free energy. Let us restrict our discussion here to the first possibility, namely dissociation into individual atoms. This will simplify the algebraic procedures of minimization and can also be shown to be the process which is in fact of first importance as the temperature is raised. The reader can check this assertion by applying the method described below to other possible processes. Our discussion concerns the reaction

$$HD \rightarrow H + D \qquad (9.34)$$

and we shall consider the temperature range $T > 100\,K$ so that the approximate rotational form $q_r \approx T/\theta_r$ can be used. It is important in expressing the various partition functions that they be all referred to the same zero, defined when the atoms are completely separated and at rest. Relative to that zero, a molecule imagined to have no rotational or vibrational energy would have a potential energy denoted by $-D_e = -7.56 \times 10^{-19}$ joules (of course the molecule must have at least the zero point energy $\tfrac{1}{2}k\theta_v$ in addition, so the dissociation energy at $0\,K$ is numerically less than D_e). Thus, from equation (5.45),

$$q_{HD} = \left(\frac{2\pi m_{HD} kT}{h^2}\right)^{3/2} g_{HD} V \left(\frac{T}{\theta_r}\right) \tfrac{1}{2}\operatorname{cosech}\left(\frac{\theta_v}{2T}\right) \exp\left(\frac{D_e}{kT}\right) \qquad (9.35)$$

Also

$$q_H = \left(\frac{2\pi m_H kT}{h^2}\right)^{3/2} g_H V \qquad (9.36)$$

and

$$q_D = \left(\frac{2\pi m_D kT}{h^2}\right)^{3/2} g_D V \qquad (9.37)$$

The nuclear spin quantum numbers of H and D are $\tfrac{1}{2}$ and 1 respectively, so that $g_H = 2$ and $g_D = 3$. For the molecule HD degeneracy (i.e., the number of possible configurations) associated with nuclear spin is simply the product $g_{HD} = g_H g_D = 6$.

If the three gases were non-reacting, the Helmholz free energy F would be given by an extension of equation (9.31), i.e.

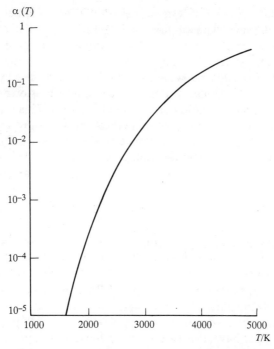

Figure 9.3 The degree of dissociation $\alpha(T)$ for HD according to equation (9.41), for an HD number density of 2.69×10^{25} molecules m^{-3} corresponding to NTP conditions.

$$F = -kT\left[N_{HD}\ln\left(\frac{eq_{HD}}{N_{HD}}\right) + N_H\ln\left(\frac{eq_H}{N_H}\right) + N_D\ln\left(\frac{eq_D}{N_D}\right)\right] \quad (9.38)$$

However, the components *do* react, so that N_{HD}, N_H and N_D are related in a way which depends on the prevailing conditions. For concreteness, suppose that initially there are N^0_{HD}/V molecules of HD in a box of volume V at temperature T, and ask what fraction α of these will dissociate as time passes. Clearly, F is now a function of α and we shall use the principle that F tends towards a minimum at equilibrium (see Section 4.3 and Section 12.1). By separate conservation of H and D atoms, we find

$$N_{HD} = (1-\alpha)N^0_{HD} \qquad N_H = \alpha N^0_{HD} \qquad N_D = \alpha N^0_{HD} \quad (9.39)$$

Hence, from (9.38),

$$F = -N^0_{HD}kT\left[(1-\alpha)\ln\left(\frac{eq_{HD}}{(1-\alpha)N^0_{HD}}\right) + \alpha\ln\left(\frac{eq_H}{\alpha N^0_{HD}}\right)\right.$$
$$\left. + \alpha\ln\left(\frac{eq_D}{\alpha N^0_{HD}}\right)\right] \quad (9.40)$$

This free energy can be minimized by differentiating F with respect to α. After rearrangement, this gives the exact result:

$$\frac{(1-\alpha)}{q_{HD}} = \frac{\alpha^2 N_{HD}^0}{q_H q_D} \qquad (9.41)$$

Thus, by using equations (9.35), (9.36) and (9.37) and specifying the number of density N_{HD}^0/V we can solve explicitly for $\alpha(T)$. Figure 9.3 shows the degree of dissociation $\alpha(T)$ versus T for a number density of 2.69×10^{25} molecules m^{-3} (chosen arbitrarily because it corresponds to NTP conditions), and at 3000 K, α is about 2 %. It is fairly common practice to specify a *pressure* (usually 1 atmosphere) rather than a number density and the reader may care to confirm the figure of 6 % quoted in Section 9.3a (see Question 9.15).

This method can be applied to many other gas reactions, and results are often alternatively expressed in terms of an equilibrium constant $K(T)$ which, for the case discussed above, might be defined as

$$K(T) \equiv \frac{(N_{HD}/V)}{(N_H/V)(N_D/V)} = \frac{1-\alpha}{\alpha^2}(V/N_{HD}^0) = \frac{2_{HD}}{2_H 2_D} \qquad (9.42)$$

It is also common to define $K_p(T)$ in terms of the ratios of partial pressures:

$$K_p(T) \equiv \frac{p_{HD}}{p_H p_D} = K(T)/kT \qquad (9.43)$$

It is important to realize that the equilibrium conditions may for many reactions take an extremely long time to arrive in the absence of a suitable catalyst. For instance, a gaseous mixture of H_2 and O_2 does not spontaneously combine to form H_2O (at least not in a measurable time scale) although a calculation like the one above would seem to favour H_2O. However, the presence of platinum as a catalyst brings about the reaction. A fuller treatment of gas reactions would have to include a study of reaction rates, which we omit here. Finally, since we have devoted attention to HD, the reader may wish to refer to a compilation[7] of thermal properties of hydrogen in its various isotopic and ortho-para modifications. However, we have chosen HD only to illustrate an approach, and the methods of this section could easily be adapted to, for instance, ionization of indeed any gas reaction such as $CO_2 + H_2 \rightarrow CO + H_2O$ or $2NH_3 \rightarrow N_2 + 3H_2$.

136 ASPECTS OF THE CLASSICAL GAS

9.6 Maxwell–Boltzmann energy distribution

We have referred to the Maxwell–Boltzmann distribution in deriving the appropriate temperature regime where the approximation (9.15) for the canonical partition function is valid. It should of course be derived from the theory and we shall do this in Section 10.2. However, we can at this stage give a less rigorous proof. The probability $p(\epsilon)$ that one particle will be found with energy ϵ was calculated in Section 6.5 with the result

$$p(\epsilon) = \frac{g \exp(-\epsilon/kT)}{q} \qquad (9.44)$$

where $g\ (= 2S + 1)$ is the spin degeneracy of the energy level and we shall take q to be the translational partition function,

$$q = gV \left(\frac{2\pi mkT}{h^2}\right)^{3/2} \qquad (9.45)$$

Given that for a gas the energy levels are close together, we can say that the probability that a molecule has an energy in the range $d\epsilon$ at ϵ is $p(\epsilon)d\rho(\epsilon)$ where $d\rho(\epsilon)$ is the number of states in the interval $d\epsilon$ and is given by equation (9.6). Now it is not unreasonable to suggest that if we take the probability of one molecule having an energy in the range of interest, and multiply it by the total number N of molecules, the product is the number of molecules in the same range. Such a procedure immediately yields equation (9.7):

$$dn = \frac{2\pi N}{(\pi kT)^{3/2}} \epsilon^{1/2} \exp\left(\frac{-\epsilon}{kT}\right) d\epsilon$$

The last and apparently innocent step in this proof does imply a different interpretation of probability than we have been using, that is, that probability may be interpreted as a frequency of occurrence of an event. Whether this is always true is a matter of debate, but since we will give a rigorous proof of equation (9.7) in Section 10.2 we do not pursue the point further here.

CHAPTER 9: SUMMARY OF MAIN POINTS

1. The canonical partition function Q for a gas at arbitrary temperature does not factorize.
2. For solids, as for example in the Einstein model, the problem does not arise because although atoms are quantum-mechanically indistinguishable, the sites which they occupy are not; in that situation there is a most

helpful theorem to the effect that $Q = q^N$ if the subsystems are all the same or $Q = \Pi q_i$ if they are different.

3. The problem in gases is less easy to dispose of with complete generality. However, a regime is identified in which $Q = q^N/N!$ to a very good approximation. The regime corresponds to a requirement that the temperature must exceed the value $\hbar^2(N/V)^{2/3}/3mk$.

4. It is shown that in practice this is not very restrictive in that in all but a tiny number of exceptional and interesting cases involving the lightest atoms, hydrogen and helium (see Chapter 11), uncondensed gases obey the 'high-temperature criterion' mentioned above and given in equation (9.22).

5. This opens the way to a use of the approximate Q to a derivation of classical gas properties.

6. The cases discussed include (9.2) the monatomic gas in which atoms are treated as capable only of translational motion, (9.3) the polyatomic gas in which the molecules have internal modes in addition to translational motion, deuterium hydride and nitric oxide being taken as examples, (9.4) non-reacting gas mixtures, and (9.5) gases which may dissociate.

7. Finally, a preliminary proof is given of the Maxwell–Boltzmann energy distribution in a classical gas.

QUESTIONS

9.1 Obtain an approximate expression for the mean free path between 'collisions' of wave packets associated with argon atoms in a gas at NTP. Hence show that the time atoms spend *between* collisions is much greater than that spent *during* collision. Though infrequent, collisions make possible the establishment of thermal equilibrium in the gas.

9.2 By using the partition function (9.30) for a mixture of two non-reacting gases, prove the results

$$C_V = \frac{3}{2}(N_A + N_B)k$$

and

$$pV = (N_A + N_B)kT$$

quoted in equations (9.32) and (9.33).

9.3 A container with a partition dividing the space into two equal volumes V is filled with N molecules of a gas in one half and N molecules of

138   ASPECTS OF THE CLASSICAL GAS

another gas in the other half. Show that the partition function for the whole system is

$$Q = \frac{q_1^N q_2^N}{N!\,N!}$$

where $q_i = V/\Lambda_i^3$ and $\Lambda_i = h/(2\pi m_i kT)^{1/2}$.

How is this expression for Q modified when the partition is removed? Using $S = kT(\partial \ln Q/\partial T)_{V,N} + k \ln Q$, calculate the 'entropy of mixing', i.e., the increase of entropy on removing the partition.

Finally, show that if the gases are indistinguishable (in the strict quantum-mechanical sense) the partition function after removal of the partition is

$$Q = \frac{q^{2N}}{(2N)!}$$

and that the entropy increase on removing the partition is then zero. (The fact that there is an entropy of mixing when the gases are distinguishable but not when they are indistinguishable was seen as paradoxical in the 19th century. The so-called 'Gibbs paradox' was resolved only when quantum mechanics outlawed the notion of 'almost indistinguishable' which led to the perception of a paradox.)

9.4   Volume V_1 contains a perfect Maxwell–Boltzmann gas made up of N atoms possessing no internal structure, at a pressure p_1 and a temperature T; volume V_2 contains the same perfect Maxwell–Boltzmann gas also of N atoms at the same temperature T but at a different pressure p_2.

If the partition is removed and the gases diffuse into each other, show that the entropy change in the diffusion process is

$$Nk \ln \frac{(p_1 + p_2)^2}{4 p_1 p_2}$$

If the initial pressures as well as the initial temperatures had been the same, then the above formula shows that the change of entropy is zero. Explain how a spontaneous process like diffusion can occur with no entropy change.

9.5   Saturated mercury vapour at the normal boiling point (630 K) is monatomic and satisfies the perfect gas laws. Calculate its molar entropy. Further, given that the latent heat of vaporization of mercury is 5.93×10^4 J mol^{-1}, use the thermodynamic relation $\Delta S = L/T$ to find the molar entropy of liquid mercury at 630 K and 1 atmosphere pressure.

9.6 We have established that in several kinds of system (including Debye solids and perfect gases) the actual energy level structure is so fine that summing over states to obtain Q may validly be replaced by integration. Let us now make an analysis of a Maxwell–Boltzmann gas of molecular permanent electric dipoles (of moment p) which in the absence of an electric field are each free to orient in any arbitrary direction. When field E is turned on, each possible orientation is associated with an energy $(-pE\cos\theta)$ which is not quantized. Let the temperature be T.

(a) Obtain a properly normalized expression for the probability $\rho(\theta)d\theta$ of a particular dipole being oriented between θ and $\theta + d\theta$ with respect to the direction of the E-field. What quantity plays the role of Q in your expression?

(b) Compute the expectation value of the component $\langle p_z \rangle$ of the dipole moment along the E-direction.

(c) Plot $\langle p_z \rangle / p$ versus pE/kT.

9.7 Following the recipe suggested by the previous question, it should now be possible to make allowance, in treating the classical gas, for the presence of the gravitational field by adding a potential energy term mgz which is not quantized. Consequently, integrals are appropriate. Consider a box of height L and cross-sectional area A.

(a) Determine the probability density $\rho(\epsilon, z)$ where $\rho(\epsilon, z)d\epsilon dz$ is the probability that an atom has kinetic energy in range $d\epsilon$ and height in range dz.

(b) Check that your new form reduces to the old when you carefully take the limit $g \to 0$.

(c) Calculate a probability distribution for height alone (by a reduction procedure), that is, for height regardless of the momentum or the 'horizontal' position.

9.8 The table below gives some results for the heat capacity of gaseous nitrogen N_2. Plot these figures and attempt to fit them to an appropriate theoretical expression derived from equation (9.29) for F. Hence find θ_v for nitrogen and deduce its characteristic frequency.

T(K)	C_V/Nk	T(K)	C_V/Nk
100	2.500	2000	3.330
500	2.558	2500	3.409
1000	2.936	3000	3.460
1500	3.194		

140 ASPECTS OF THE CLASSICAL GAS

9.9 The reason why a discussion of H_2 was avoided in the text is that it can exist in two subtly distinct forms having behaviour differences which, though trivial at high temperatures, are magnified at low temperatures. The difference is simply that the spins of the two nuclei may be parallel (ortho-hydrogen) or antiparallel (para-hydrogen). This has implications for the symmetries of the appropriate eigenfunctions.

What is the equilibrium ratio of ortho- to para-hydrogen at 30 K and at the limit of high T? You will need to bear in mind the following facts:

(a) The distance between protons in H_2 is 0.0747 nm.

(b) Only rotational energies need to be taken into account because other modes are unaffected by symmetry.

(c) The spin degeneracies g associated with ortho- and para-hydrogen are 3 and 1 respectively. (Why ?)

(d) The rotational energies are:

$$\epsilon_J = \frac{\hbar^2 J(J+1)}{2I} \text{ where } J = 1, 3, 5 \ldots \text{ for para-}H_2$$
$$\text{and } J = 0, 2, 4 \ldots \text{ for ortho-}H_2$$

(e) Classical statistics are valid for H_2 molecules at the temperatures considered.

9.10 Calculate an expression for the heat capacity of a gas of atoms, each of which possesses both translational and electronic energies, where the electronic partition function is of the form

$$q_e = g_0 + g_1 \exp\left(-\frac{\epsilon_1}{kT}\right)$$

where g_0 and g_1 are degeneracies.

Show that at high and low temperatures the electronic contributions fall to zero. Why does this happen?

9.11 The molecules of a diatomic gas have energy levels

$$\epsilon_J = \frac{\hbar^2 J(J+1)}{2I}$$

where $J = 0, 1, 2 \ldots$ and I is the moment of inertia of the molecules. Each level is $(2J + 1)$-fold degenerate. For carbon monoxide (CO) $I = 1.3 \times 10^{-46}$ kg m². What is the heat capacity of carbon monoxide at 300 K?

9.12 At a given temperature and volume a gas of AB particles is 50 % dissociated into A particles and B particles. The pressure is p_0, the volume is then doubled, the temperature remaining constant. The new pressure is p_1. Show that

$$\frac{p_1}{p_0} = \frac{1 + \sqrt{5}}{6}$$

9.13 At very high temperatures, atomic hydrogen ionizes to form one electron and one proton, with an ionization energy of 2.17×10^{-18} J. Assuming the electron and proton gases to behave classically, find the temperature at which 1 % dissociation has occurred in a volume containing 10^{25} atoms m^{-3}. Is your calculated temperature sufficiently high to justify the assumption of classical behaviour?

9.14 A plasma (that is, a gas containing an appreciable number of dissociated positive and negative charges) is more easily made with caesium vapour than with hydrogen because the ionization energy is only 6.22×10^{-19} J. Caesium has atomic weight 133. As g_{Cs} does not affect the result strongly you may assume g_{Cs} is unity.

(a) Express the degree of dissociation, α, of caesium vapour in terms of temperature and *pressure*.

(b) Evaluate your expression for $T = 1200$ K and $p = 100$ Pa ($\sim 10^{-3}$ atm).

9.15 Confirm, by solving equation (9.41), that the degree of dissociation in HD at 3000 K and 1 atmosphere pressure is a little under 6 %.

9.16 The inner part of the sun consists of very hot dense gases, while its outer corona is cooler and less dense. Studies of spectral lines of light from the sun indicate that atoms may be ionized in the corona, and yet not be ionized in regions closer to the sun where the absolute temperature is much higher. How do you explain these observations?

9.17 A particle of mass m in a one-dimensional well of length l has non-degenerate energy

$$\epsilon = \frac{h^2 n^2}{8ml^2}$$

where n is a non-zero integer. Prove that the density of states, $g(\epsilon)$, for energies ϵ much greater than $h^2/8ml^2$ is given by

$$g(\epsilon) = \frac{2ml}{h(2m\epsilon)^{1/2}}$$

Evaluate the heat capacity at high temperatures of N non-interacting particles obeying Boltzmann statistics in this potential well. Why would you expect your expression for the heat capacity to fall when $kT \ll h^2/8ml^2$?

9.18 An ethane (C_2H_6) molecule has an axial C−C bond and each carbon atom has a symmetrical trio of hydrogen atoms attached to it. The two CH_3 groups are thus like pyramids with an equilateral triangular base of hydrogen atoms and a carbon atom at the apex. The two groups can rotate about the C−C axis in opposite directions so that the molecule becomes twisted. This mode is thus distinguishable from the one in which the molecule rotates bodily about the C−C axis.

Why would you expect the potential energy of the molecule to be periodic (with period $2\pi/3$) in the relative angular displacement between the two groups?

Discuss qualitatively the dependence of the heat capacity of this mode on temperature and make specific quantitative predictions for its form at (a) sufficiently high T, and (b) sufficiently low T. Explain in each case what you mean by 'sufficiently'.

9.19 Explain the conditions for the canonical partition function q of a single homonuclear diatomic molecule to be expressible as the product $q_{trans} \times q_{vib} \times q_{rot} \times \ldots$ where q_{trans}, q_{vib}, and q_{rot} refer to translational, vibrational, and rotational motions respectively, where the product may also contain factors representing other possible properties such as magnetic, electric, or nuclear. Write down and explain the detailed forms of q_{trans}, q_{vib}, and q_{rot}, but without attempting to evaluate the sums. Explain why only q_{trans} contains the volume V. Assuming further that q_{trans} is directly proportional to V, prove that for a single homonuclear diatomic molecule, the pressure is given by $pV = kT$.

Consider a gas of homonuclear diatomic molecules some of which have dissociated according to $A_2 \leftrightarrows A + A$. Explain why, if there are N_1 atoms of type A each with canonical partition function q_1, and N_2 molecules of type A_2 each with canonical partition function q_2, the canonical partition function Q of the gas is

$$Q = \frac{q_1^{N_1} q_2^{N_2}}{N_1! N_2!}$$

For the following calculation you may treat both N_1 and N_2 as numerically large and continuously variable. By minimizing the free energy $(-kT \ln Q)$ with respect to variations in N_1 at constant T, subject to the condition that the total number of atoms ($N_1 + 2N_2$) be constant, show that at equilibrium

$$N_1^2/N_2 = q_1^2/q_2$$

REFERENCES

1. For example, Estermann, Simpson and Stern, *Phys. Rev.* (1947), **71**, 238.
2. Readers wishing to read more about this topic would do well to turn to Greytak T.J. and Kleppner D., *Lectures on Spin-polarized Hydrogen*, (North-Holland, 1984), reprinted from Les Houches Session XXXVIII: New Trends in Atomic Physics, 28 June–29 July 1982.
3. Clusius and Bartholomé, *Zeit. für Elektrochem.* (1934), **40**, 524.
4. Johnston and Long, *J. Chem. Phys.* (1934), **2**, 389, with a minor numerical correction in *J. Chem. Phys.* (1934), **2**, 710.
5. Johnston and Chapman, *J. Amer. Chem. Soc.* (1933), **55**, 153; Witner, *J. Amer. Chem. Soc.* (1934), **56**, 2229.
6. Eucken and d'Or, *Gött. Nachr.* (1932), 107.
7. Wooley, Scott and Brickwedde, *J. Res. Nat. Bur. Standards* (1948), **41**, 379.

CHAPTER 10

Gases in the Grand Canonical Distribution

10.1 General formulation
10.2 The classical gas and the Maxwell–Boltzmann distribution
10.3 The Fermi gas
10.4 Fluctuations in a fermion system
10.5 The Bose gas
10.6 Fluctuations in Bose systems
Questions
References

10.1 General formulation

We saw in the last chapter that, although it is possible to write down the canonical partition function in the form of a summation for a gas, it is not possible to perform the summation explicitly. A very good approximation exists, however, provided that $T \gg \theta_0$; this is given by equation (9.15). We were able to argue that this criterion was satisfied by all real gases at equilibrium (spin-polarized monatomic hydrogen, deuterium, and tritium might prove to be exceptions although the experimental difficulties involved in investigating them are formidable). This is due to the fact that the separation between atoms or molecules in real gases is large compared with the range of interaction and the de Broglie wavelength. For real gases, therefore, we have a perfectly serviceable explicit form for the canonical partition function. However, we cannot let the matter rest there because there are a number of 'gas-like' systems for which the criterion $T \ll \theta_0$ is not satisfied. Examples of such systems include (a) the conduction electrons in metals and semiconductors, (b) photons, (c) liquid helium four, (d) liquid helium three, and perhaps (e) the interiors of neutron or white dwarf stars.

GENERAL FORMULATION

The difficulty in the exact evaluation of the canonical partition function arises from the restriction on the sets of occupation numbers $\{n_s\}$ that they must satisfy $\Sigma_s n_s = N$. This restriction will be removed if we go to the grand canonical distribution. For this distribution (see equation (4.55)), we first sum over all quantum mechanical states for a fixed N, and then over all N from zero to infinity. For the case of identical particles this can be done by first summing over all sets $\{n_s\}$ consistent with $\Sigma_s n_s = N$, and then over all N. These two steps can of course be combined by summing over all sets $\{n_s\}$ *without* the restriction $\Sigma_s n_s = N$. Thus from equations (4.55), (8.18) and (8.23)

$$\Xi = \sum_{N=0}^{\infty} \left[\exp\left(\frac{\mu N}{kT}\right) \sum_{\{n_s\}}^{X} \exp\left(-\frac{E_{\{n_s\}}}{kT}\right) \right] \qquad (10.1)$$

$$= \sum_{\{n_s\}}^{X} \exp\left[\sum_s \frac{\mu n_s - \epsilon_s n_s}{kT} \right] \qquad (10.2)$$

The last step simply combines the summations (removing the restriction on the sets $\{n_s\}$) and uses $N = \Sigma_s n_s$ and $E_{\{n_s\}} = \Sigma_s \epsilon_s n_s$. The next step is to observe that the exponential of a sum is equal to the product of exponentials i.e., $\exp(\Sigma_s x_s) = \Pi_s \exp(x_s)$, so that

$$\exp\left[\sum_s \frac{\mu n_s - \epsilon_s n_s}{kT} \right] \equiv \prod_s \exp\left[\frac{\mu n_s - \epsilon_s n_s}{kT} \right] \qquad (10.3)$$

Finally, since the n_s now appear quite independently in (10.3), the notation $\{n_s\}$ in (10.2) may be dropped in favour of n_s which may run over all allowed values from 0 to X ($X = 1$ for fermions; $X = \infty$ for bosons (the artificial case of intermediate statistics, with arbitrary X, is the topic of question 10.1)). Thus

$$\Xi = \prod_s \sum_{n_s=0}^{X} \exp\left[\frac{\mu n_s - \epsilon_s n_s}{kT} \right] \qquad (10.4)$$

To proceed further we must specify whether we are dealing with fermions or bosons. For fermions, $X = 1$, and hence each sum over n_s consists of only two terms as

$$\Xi_f = \prod_s \left\{ 1 + \exp\left[\frac{\mu - \epsilon_s}{kT}\right] \right\}$$

i.e.

$$\ln \Xi_f = \sum_s \ln\left\{ 1 + \exp\left[\frac{\mu - \epsilon_s}{kT}\right] \right\} \qquad (10.5)$$

For bosons $X = \infty$ and each sum over n_s consists of an infinite geometric series, which is explicitly summable:

$$\sum_{n_s=0}^{\infty} \exp\left[\frac{\mu n_s - \epsilon_s n_s}{kT}\right] = \left\{1 - \exp\left[\frac{\mu - \epsilon_s}{kT}\right]\right\}^{-1} \qquad (10.6)$$

Thus

$$\Xi_b = \prod_s \left\{1 - \exp\left[\frac{\mu - \epsilon_s}{kT}\right]\right\}^{-1}$$

i.e.,

$$\ln \Xi_b = -\sum_s \ln\left\{1 - \exp\left[\frac{\mu - \epsilon_s}{kT}\right]\right\} \qquad (10.7)$$

Equations (10.5) and (10.7) may be combined in one expression:

$$\ln \Xi_{f/b} = \pm \sum_s \ln\left\{1 \pm \exp\left[\frac{\mu - \epsilon_s}{kT}\right]\right\} \qquad (10.8)$$

where the upper sign is for fermions, and the lower for bosons. The bridge equation for the grand canonical distribution (4.69) is thus

$$Z = kT \ln \Xi = \pm kT \sum_s \ln\left[1 \pm \exp\left(\frac{\mu - \epsilon_s}{kT}\right)\right] \qquad (10.9)$$

The other thermodynamic properties follow from equations (4.70) to (4.73); in particular,

$$\mathcal{N} = \left(\frac{\partial Z}{\partial \mu}\right)_{T,V} = \sum_s \left[\exp\left(\frac{\epsilon_s - \mu}{kT}\right) \pm 1\right]^{-1} \qquad (10.10)$$

This equation has a rather elegant physical interpretation. Each term in the sum is a positive number and the sum goes over all possible states. Thus we can interpret the summand as the mean number of particles occupying the single particle state s, that is, we can write

$$\mathcal{N} = \sum_s \langle n_s \rangle$$

where the mean number of particles in the sth state is

$$\langle n_s \rangle = \left[\exp\left(\frac{\epsilon_s - \mu}{kT}\right) \pm 1 \right]^{-1} \qquad (10.11)$$

Again the upper sign is for fermions and the lower for bosons. It is easy to see that for fermions the mean occupation number of the sth level is always less than 1. Since the only two possible values of n_s are 0 and 1, this is the upper bound for n_s we would expect. No such upper bound exists for the mean occupation number for bosons.

The entropy can be calculated from equation (4.71):

$$\begin{aligned} S &= \left(\frac{\partial Z}{\partial T}\right)_{\mu,V} \\ &= k\sum_s \left(\frac{\epsilon_s - \mu}{kT}\right) \left[\exp\left(\frac{\epsilon_s - \mu}{kT}\right) \pm 1\right]^{-1} \\ &\quad \pm k\sum_s \ln\left[1 \pm \exp\left(\frac{\mu - \epsilon_s}{kT}\right)\right] \end{aligned} \qquad (10.12)$$

Now by using equation (10.11), we can simplify by expressing the entropy conveniently in terms of $\langle n_s \rangle$:

$$S = -k\sum_s \left[\langle n_s \rangle \ln \langle n_s \rangle \mp (1 \pm \langle n_s \rangle) \ln (1 \mp \langle n_s \rangle) \right] \qquad (10.13)$$

We will interpret this equation separately in the sections dealing with fermions and bosons. Meanwhile, we can use equation (4.73) to arrive at a transparently simple expression for the internal energy. This requires only the substitution of equations (10.9), (10.11) and (10.12) into equation (4.73); there is some algebraic manipulation which the reader may care to try. The result, for fermions and bosons, is

$$U = \sum_s \langle n_s \rangle \epsilon_s \qquad (10.14)$$

which has an obvious physical interpretation, namely that the total energy is the sum of the energies of the mean number of particles occupying each single-particle state, times the energy of that state. Next we calculate the pressure:

$$p = \left(\frac{\partial Z}{\partial V}\right)_{\mu,T} = -\sum_s \left[\exp\left(\frac{\epsilon_s - \mu}{kT}\right) \pm 1\right]^{-1} \frac{d\epsilon_s}{dV} = -\sum_s \langle n_s \rangle \frac{d\epsilon_s}{dV} \qquad (10.15)$$

For most realistic models, ϵ_s is proportional to $V^{-\alpha}$, and then we get

$$pV = \alpha \sum_s \langle n_s \rangle \epsilon_s = \alpha U \tag{10.16}$$

where the final step makes use of equation (10.14).

We now return to evaluating the sum over single-particle states in equations (10.9) to (10.12). We look for a means of converting sums to integrals, and we shall first deal briefly with the high-temperature, low-density case as a guide to procedures for fuller treatments of Fermi and Bose gases in later sections.

10.2 The classical gas and the Maxwell–Boltzmann distribution

We found in Chapter 9 an expression for the number $d\rho$ of particle-in-a-box states with energies in the range $d\epsilon$. The quantity $g(\epsilon) \equiv d\rho/d\epsilon$ is known as the density of states and, for our gas,

$$\left. \begin{array}{ll} g(\epsilon) = \dfrac{2\pi V g}{h^3} (2m)^{3/2} \epsilon^{1/2} & (\epsilon > 0) \\ g(\epsilon) = 0 & (\epsilon < 0) \end{array} \right\} \tag{10.17}$$

This result arises from counting the possible number of states of a particle-in-a-box and has nothing whatsoever to do with what particular distribution we choose to work with. It will therefore be used throughout this chapter. However, we must remember that strictly it is an approximation derived on the assumption that ϵ can validly be treated as a quasi-continuous variable, which is most accurate for high energy values. We might say that it is true for energies $\epsilon \gg k\theta_t$. For the classical (and Fermi) gas, we can afford to be wrong about the density of low-lying states because no state is occupied by more than one atom. Consequently, at *any* temperature, the vast majority of atoms are in states for which $\epsilon \gg k\theta_t$ and equation (10.17) is valid. We shall later be unable to make a similar assertion for Bose gases in which low-lying states are of dominating importance at low temperatures. However, for present purposes we use equation (10.17) and bear in mind that the energy levels are very close together. This follows from the form of the energy levels, equation (5.31)

$$\epsilon = k\theta_t (n_x^2 + n_y^2 + n_z^2) \tag{10.18}$$

Thus

$$\frac{\Delta\epsilon}{\epsilon} = \frac{n_x^2 + n_y^2 + (n_z+1)^2 - n_x^2 - n_y^2 - n_z^2}{n_x^2 + n_y^2 + n_z^2}$$

$$= \frac{2n_z + 1}{n_x^2 + n_y^2 + n_z^2} \qquad (10.19)$$

This expression measures the closeness of the energy levels and for large n_x, n_y, n_z this means that they are very close, in which case all the sums over states may be converted immediately to integrals. In particular the expression (10.10) for $\mathcal{N}$ becomes

$$\mathcal{N} = \sum_s \left\{ \exp\left[\frac{\epsilon_s - \mu}{kT}\right] \pm 1 \right\}^{-1} \qquad (10.20)$$

$$= \int_{-\infty}^{\infty} g(\epsilon) \left\{ \exp\left[\frac{\epsilon - \mu}{kT}\right] \pm 1 \right\}^{-1} d\epsilon \qquad (10.21)$$

$$= \frac{2\pi Vg}{h^3}(2m)^{3/2} \int_0^{\infty} \epsilon^{1/2} \left\{ \exp\left[\frac{\epsilon - \mu}{kT}\right] \pm 1 \right\}^{-1} d\epsilon \qquad (10.22)$$

The lower limit of integration ($\epsilon = -\infty$) in (10.21) is included for completeness, since on occasions we might wish to allow negative energies in such cases as a gas in a potential well. The lower limit in equation (10.22) applies to the special case where the potential energy inside the box is taken as zero.

This equation determines $\mu(T)$. At high temperatures ($T \gg \theta_0$) we must expect the distinction between the Fermi and Bose gases to disappear since they are then in the regime discussed in the last chapter. Now the only algebraic difference is the ± 1. Assuming for the moment, therefore, that the ± 1 can be neglected in comparison with the exponential, we write equation (10.22) in the simplified form for $T \gg \theta_0$:

$$\mathcal{N} = \int_{-\infty}^{\infty} g(\epsilon) \exp\left[\frac{\mu - \epsilon}{kT}\right] d\epsilon \qquad (10.23)$$

$$= \frac{2\pi Vg}{h^3}(2m)^{3/2} \exp\left(\frac{\mu}{kT}\right) \int_{-\infty}^{\infty} \epsilon^{1/2} \exp\left(-\frac{\epsilon}{kT}\right) d\epsilon \qquad (10.24)$$

The integral can be performed explicitly and after rearrangement we obtain

$$\exp\left(\frac{-\mu}{kT}\right) = \left(\frac{2\pi mkT}{h^2}\right)^{3/2} \frac{gV}{N} = \left(\frac{T}{\theta_0}\right)^{3/2} \qquad (10.25)$$

or

$$\mu = -\tfrac{3}{2} k\theta_0 \left(\frac{T}{\theta_0}\right) \ln\left(\frac{T}{\theta_0}\right) \qquad (10.26)$$

The last step follows on substituting for θ_0 from equation (9.10) and it is plain that in the regime $T \gg \theta_0$, μ is always a negative quantity. We should now check that equation (10.26) is consistent with our assumption that the ± 1 could be neglected when $T \gg \theta_0$ in (10.21) or (10.22). Clearly, since all ϵ are positive, $\exp(\epsilon/kT) > 1$ for all ϵ and T. Moreover, from equation (10.26), $\exp(-\mu/kT) \gg 1$ when $T \gg \theta_0$. Hence (10.26) is consistent with $\exp[(\epsilon - \mu)/kT] \gg 1$, which was our assumption.

Finally, we are in a position to derive the Maxwell–Boltzmann distribution in a logically consistent manner (see Section 9.6 for a less satisfactory approach). If we write equation (10.24) in the form

$$\mathcal{N} = \int_0^\infty \frac{d\mathcal{N}}{d\epsilon} d\epsilon \qquad (10.27)$$

then we can interpret $(d\mathcal{N}/d\epsilon)d\epsilon = d\mathcal{N}$ as the number of particles having energies between ϵ and $\epsilon + d\epsilon$. Comparing equation (10.27) with equation (10.24) and using equation (10.25) gives, for $d\mathcal{N}$,

$$d\mathcal{N} = \frac{2\pi \mathcal{N}}{(\pi kT)^{3/2}} \epsilon^{1/2} \exp\left(-\frac{\epsilon}{kT}\right) d\epsilon \qquad (10.28)$$

which is the required result. This expression has been well verified experimentally.[1]

10.3 The Fermi gas

The results of the above section are encouraging in that they have enabled us to derive the well-known Maxwell–Boltzmann distribution. It is perhaps of even more interest to look at the quantum effects which occur at temperatures below θ_0. In this case the ± 1, which we neglected in considering the Maxwell–Boltzmann gas, now becomes important and we must consider the cases of the two different signs separately. In this section we consider the case of a gas obeying Fermi–Dirac statistics or, for short, the 'Fermi gas'. The general results that we derived in Section 10.1 apply but we are left with evaluating the formulae, which are sums over all the quantum single-particle states. In the case of the Maxwell–Boltzmann gas we replaced the sum by an integral, using the density of states (10.17), and it is necessary to ask whether we can do the same for the Fermi gas. The conversion from sum to integral is a valid approximation as we have shown for single-particle states of high energy but will get increasingly worse the lower the energy of the states.

THE FERMI GAS 151

However, for the particles of a Fermi gas, or fermions, we can at the most have only two particles in any translational state (one for each spin direction, assuming we are dealing with spin-$\frac{1}{2}$ particles). Thus, the low-lying states will have only a few particles in them compared to the macroscopic number of particles in the gas. Any error we make in calculating their contribution to the thermodynamic properties will be negligible. Thus, equation (10.10), which is an equation to determine the quantity μ, can be written

$$\mathcal{N} = \int_0^\infty \left[\exp\left(\frac{\epsilon - \mu}{kT}\right) + 1 \right]^{-1} g(\epsilon) d\epsilon \qquad (10.29)$$

or substituting for the density of states (10.17),

$$\mathcal{N} = \frac{2\pi Vg(2m)^{3/2}}{h^3} \int_0^\infty \epsilon^{1/2} \left[\exp\left(\frac{\epsilon - \mu}{kT}\right) + 1 \right]^{-1} d\epsilon \qquad (10.30)$$

This then is the equation that has to be solved for μ. An elegant technique has been found by Leonard,[2] or alternatively results can be obtained numerically using an iterative method (see Appendix 4 for some tabulations). We can however find the low-temperature form very easily. Let the resulting value of μ in the limiting case $T \to 0$ be denoted by μ_0. Then, in the integral, for those values of ϵ which are greater than μ_0, the integrand tends to zero as $T \to 0$; whereas for those values of ϵ which are less than μ_0, the integrand tends to $\epsilon^{1/2}$. Hence the upper limit of the integral may be taken as μ_0 and the integral reduces to

$$\mathcal{N} = \frac{2\pi Vg(2m)^{3/2}}{h^3} \int_0^{\mu_0} \epsilon^{1/2} d\epsilon \qquad (10.31)$$

so that, on rearrangement,

$$\mu_0 = \frac{h^2}{8m} \left(\frac{6\mathcal{N}}{\pi Vg}\right)^{2/3} \qquad (10.32)$$

This zero-temperature value of μ_0 is known as the 'Fermi energy' and the characteristic temperature defined as

$$T_f = \frac{\mu_0}{k} = \frac{h^2}{8mk} \left(\frac{6\mathcal{N}}{\pi Vg}\right)^{2/3} \qquad (10.33)$$

is called the 'Fermi degeneracy temperature'. The reader will notice on comparing (10.33) with (9.10) that T_f is very close to θ_0, so that the quantum region, or the region of Fermi degeneracy, is that for which $T \ll \theta_0$. Figure 10.1 shows the chemical potential for a Fermi gas as a function of

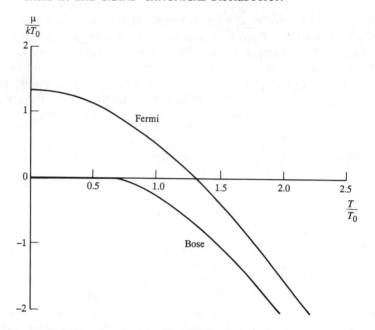

Figure 10.1 The chemical potentials of Fermi and Bose gases, plotted in terms of a convenient common temperature as used in Appendix 4:

$$T_0 \equiv \frac{h^2}{2mk}\left(\frac{\mathcal{N}}{2\pi Vg}\right)^{2/3}$$

temperature. It shows that for temperatures greater than about $1.3\, T_0$, the chemical potential becomes negative and its dependence becomes increasingly classical (see Section 10.2 above).

The summation in equation (10.14) can be replaced by an integral to give for the internal energy

$$U = \frac{2\pi Vg(2m)^{3/2}}{h^3}\int_0^\infty \epsilon^{3/2}\left[\exp\left(\frac{\epsilon-\mu}{kT}\right)+1\right]^{-1}d\epsilon \tag{10.34}$$

Equation (9.3) shows that the translational energies depend upon the volume as $V^{-2/3}$ so that $\alpha = \frac{2}{3}$ and

$$pV = \tfrac{2}{3}U \tag{10.35}$$

In the low-temperature limit the integral in (10.34) can be evaluated in a similar manner to that of (10.32) and the result is

$$U = \tfrac{3}{5}\mathcal{N}\mu_0 \tag{10.36}$$

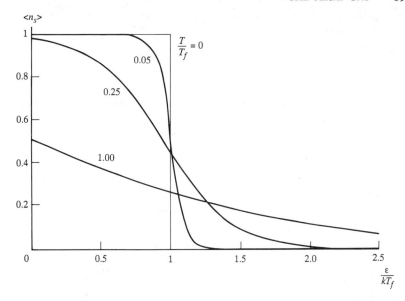

Figure 10.2 The behaviour of the expectation value of the Fermi occupation number $\langle n_s \rangle$ as a function of ϵ/kT_f from equation (10.12), with the + sign and with μ obtained by solving equation (10.29) for a few values of T/T_f. At low T/T_f there is a fairly sharp step from $\langle n_s \rangle = 1$ to $\langle n_s \rangle = 0$, but when $T/T_f \gtrsim 1$ the variation is smooth.

so that

$$pV = \tfrac{2}{5} \mathcal{N} \mu_0 \tag{10.37}$$

The reason why the internal energy and the pressure are not zero at zero temperature (as they are in the classical gas), is entirely a consequence of the Fermi exclusion principle. Since all the particles cannot be put into the single-particle ground state, we have to put two into each state, which gradually increases the energy, until all the particles are used up, the most energetic having an energy equal to the Fermi energy. Figure 10.2 shows $\langle n_s \rangle$ as a function of ϵ_s/kT using equation (10.11) (with the plus sign for fermions), for various values of T. It shows that at $T = 0$ all single-particle states are occupied up to the Fermi energy $\mu_0 = kT_f$ and above that energy the single-particle states are unoccupied.

To obtain the specific heat we need to calculate the temperature dependence of both the chemical potential and the internal energy. This is done in Appendix 5, and the result for U is

$$U = \tfrac{3}{5} \mathcal{N} \mu_0 \left[1 + \frac{5\pi^2}{12} \left(\frac{T}{T_f} \right)^2 \cdots \right] \tag{10.38}$$

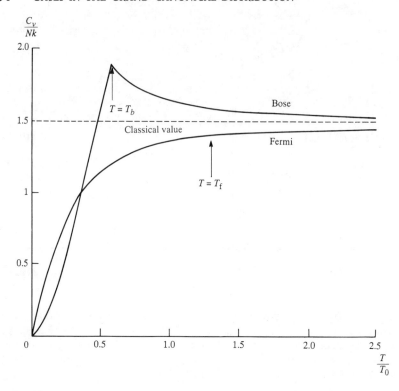

Figure 10.3 The heat capacities of Fermi and Bose gases, again plotted as functions of T/T_0 where

$$T_0 \equiv \frac{h^2}{2mk}\left(\frac{\mathcal{N}}{2\pi V g}\right)^{2/3}$$

The Bose gas shows a cusp at $T = T_b$, and at low T, $C_V \propto T^{3/2}$. The Fermi gas shows smooth behaviour with $C_V \propto T$ at low T. For $T/T_0 \gtrsim 2.5$ both varieties approximate to classical behaviour.

so that, on differentiation with respect to temperature,

$$C_V = \left(\frac{\partial U}{\partial T}\right)_V = \tfrac{1}{2}\pi^2 \mathcal{N} k \left(\frac{T}{T_f}\right)\ldots \qquad (10.39)$$

The low-temperature specific heat is thus linear. Figure 10.3 shows the variation of the heat capacity with temperature. It is also possible to obtain high-temperature corrections to the classical value of $\tfrac{3}{2}\mathcal{N}k$. This is the subject of Question 10.2.

Finally in this section we obtain an interpretation of the expression (10.13) for the entropy. Taking always the upper signs since we are dealing with fermions, we have

$$S = -k\sum_s [\langle n_s \rangle \ln (\langle n_s \rangle) + (1 - \langle n_s \rangle) \ln (1 - \langle n_s \rangle)] \quad (10.40)$$

Let us denote by P_s^0 the probability that single-particle state s is unoccupied, and by P_s^1 the probability that it *is* occupied. Then by definition of $\langle n_s \rangle$,

$$(0 \times P_s^0) + (1 \times P_s^1) = \langle n_s \rangle \quad (10.41)$$

and

$$P_s^0 + P_s^1 = 1 \quad (10.42)$$

Thus

$$P_s^1 = \langle n_s \rangle \quad (10.43)$$

and

$$P_s^0 = 1 - \langle n_s \rangle \quad (10.44)$$

Consequently equation (10.40) may be written

$$S = -k \sum_s \left[P_s^0 \ln P_s^0 + P_s^1 \ln P_s^1 \right] \quad (10.45)$$

Thus the total uncertainty S/k is the sum of the uncertainties for the occupancies of the single-particle states. Since whether a particular state is occupied or not is independent of the occupancy of the other states, and since under these conditions the uncertainties are additive, equation (10.45) is just the result we would expect.

10.4 Fluctuations in a fermion system

Finally, in this discussion of a collection of non-interacting fermions we turn to the question of fractional deviations or fluctuations. First let us consider the fluctuation in the occupation of a single level with quantum number s. This is easy to calculate since we have already found the probabilities for the possible occupancies of a particular state. Thus from equations (10.43) and (10.44) we have

$$\langle n_s^2 \rangle = P_s^0 \times (0)^2 + P_s^1 \times (1)^2 = \langle n_s \rangle \quad (10.46)$$

Thus

$$\frac{\langle \Delta n_s \rangle^2_{rms}}{\langle n_s \rangle^2} \equiv \frac{\langle n_s^2 \rangle - \langle n_s \rangle^2}{\langle n_s \rangle^2} = \frac{1 - \langle n_s \rangle}{\langle n_s \rangle} \qquad (10.47)$$

In general, this is clearly not vanishingly small and may have any value in the range 0 to ∞ depending on $\langle n_s \rangle$, and it is equal to 1 for $\langle n_s \rangle = \frac{1}{2}$. We now have to satisfy ourselves that in spite of this result for a single state s, the fluctuation in the total number of particles is vanishingly small. This is argued as follows:

$$\langle N \rangle = \mathcal{N} = \sum_s \langle n_s \rangle \qquad (10.48)$$

Also

$$\langle N^2 \rangle = \sum_s \sum_r \langle n_s n_r \rangle \qquad (10.49)$$

Hence

$$\frac{\langle \Delta N \rangle^2_{rms}}{\mathcal{N}^2} = \frac{1}{\mathcal{N}^2} \left[\sum_s \sum_r \langle n_s n_r \rangle - \sum_s \sum_r (\langle n_s \rangle \langle n_r \rangle) \right] \qquad (10.50)$$

Now we make a distinction between terms with $s = r$ and terms with $s \neq r$, and express equation (10.50) as

$$\frac{\langle \Delta N \rangle^2_{rms}}{\mathcal{N}^2} = \frac{1}{\mathcal{N}^2} \sum_s \left[\langle n_s^2 \rangle - \langle n_s \rangle^2 \right] + \frac{1}{\mathcal{N}^2} \sum_s \sum_{\substack{r \\ s \neq r}} \left[\langle n_s n_r \rangle - \langle n_s \rangle \langle n_r \rangle \right] \qquad (10.51)$$

Now since the particles are independent, the occupancies of *different* levels are independent and hence, for $s \neq r$, $\langle n_s n_r \rangle = \langle n_s \rangle \langle n_r \rangle$ and the second term in (10.51) vanishes identically. Thus, and remembering that $\langle n_s^2 \rangle = \langle n_s \rangle$ from equation (10.46), we get

$$\frac{\langle \Delta N \rangle^2_{rms}}{\mathcal{N}^2} = \frac{1}{\mathcal{N}^2} \sum_s \langle n_s \rangle (1 - \langle n_s \rangle) \qquad (10.52)$$

$$= \frac{1}{\mathcal{N}} - \frac{1}{\mathcal{N}^2} \sum_s \langle n_s \rangle^2 \qquad (10.53)$$

But all the $\langle n_s \rangle$ lie between 0 and 1 so that $\Sigma_s \langle n_s \rangle^2 < \mathcal{N}$, and

$$\frac{\langle \Delta N \rangle^2_{\text{rms}}}{\mathcal{N}^2} \sim \frac{1}{\mathcal{N}} \qquad (10.54)$$

as expected.

10.5 The Bose gas

The Bose gas is a very good example of how nature can surprise us. Just when we think we have a rigorous procedure that will work in all situations, we find an exception to the rule. In Section 10.1, we gave a general theory applicable to the classical gas, the Fermi gas and the Bose gas. For the first two cases we made the reasonable and correct step of converting sums to integrals, so that we could evaluate the formulae, or at least get low- and high-temperature approximations. For the Bose gas such a procedure would be incorrect, because of a phenomenon known as Bose condensation or macroscopic occupation of the ground state. This phenomenon arises as follows: consider equation (10.10) for the Bose case by selecting the minus sign so that

$$\mathcal{N} = \sum_s \left[\exp\left(\frac{\epsilon_s - \mu}{kT}\right) - 1 \right]^{-1} \qquad (10.55)$$

If the summand is to remain positive as $T \to 0$ as it must (since its physical interpretation is that it is the mean occupation number of the sth level), then $\mu < \epsilon_s$ for all s. In particular $\mu < \epsilon_0$ where ϵ_0 is the single-particle ground state energy. ϵ_0 is thus the value of μ at $T = 0$ and at that temperature all other occupation numbers must be zero. We note that the presence in one single-particle quantum state of a macroscopic number of particles is a rare phenomenon. However, we return to considering equation (10.55) at non-zero temperatures. This equation is the one we have to solve for a given $\mathcal{N}$ to find the chemical potential μ. Now we know from the considerations of Section 10.2 that μ has to be negative for $T \gg \theta_0$ for bosons as well as fermions. But now for bosons as $T \to 0$, μ has to approach ϵ_0, which is miniscule but positive. A consequence of this is that at some finite temperature T_b, μ has to be zero. To clarify the situation we anticipate the answer and show μ as a function of temperature in Figure 10.1. Assuming for the moment that the conversion to an integral is valid at this value of μ, equation (10.55) can be written

$$\mathcal{N} = \frac{2\pi g V}{h^3} (2m)^{3/2} \int_0^\infty \epsilon^{1/2} \left[\exp\left(\frac{\epsilon}{kT_b}\right) - 1 \right]^{-1} d\epsilon \qquad (10.56)$$

$$= \frac{2\pi g V}{h^3} (2m)^{3/2} (kT_b)^{3/2} \int_0^\infty x^{1/2} [\exp(x) - 1]^{-1} dx \qquad (10.57)$$

The definite integral is the Riemann zeta function $\zeta(3/2)$ where, in general

$$\zeta(n) = \frac{1}{\Gamma(n)} \int_0^\infty \frac{x^{n-1}}{e^x - 1} \, dx \tag{10.58}$$

$\zeta(3/2)$ has the numerical value 2.612, so that after rearrangement we get from equation (10.57),

$$T_b = \frac{h^2}{2\pi mk} \left(\frac{\mathcal{N}}{2.612 g V} \right)^{2/3} = \frac{\theta_0}{(2.612)^{2/3}} \tag{10.59}$$

which is very close to θ_0. So for $T \gg T_b$ we are in the classical gas regime. In this section we are interested in the opposite regime $T \ll T_b$. T_b is known as the Bose degeneracy temperature. Before going further we have to check that our procedure in going from the sum to the integral in calculating T_b is valid. If the procedure is correct then we need to show that the mean occupation numbers for the states with low energies are small. This can be done by evaluating $\langle n_s \rangle$ for a few low-lying states at T_b. The energies will be given by equation (5.31) with appropriate choices of n_x, n_y, and n_z, and we find the first few energy levels to be ϵ_0, $2\epsilon_0$, $3\epsilon_0$, $11\epsilon_0/3$, and $4\epsilon_0$ where

$$\epsilon_0 = \frac{3h^2}{8mV^{3/2}} = \frac{3\pi}{4} \left(\frac{2.612 g}{\mathcal{N}} \right)^{2/3} kT_b \tag{10.60}$$

so that $\epsilon_0 \ll kT_b$ for macroscopic $\mathcal{N}$. Thus, at $T = T_b$ (where $\mu = 0$ by definition)

$$\langle n_0 \rangle = \left[\exp\left(\frac{\epsilon_0}{kT_b} \right) - 1 \right]^{-1}$$

$$\approx \frac{kT_b}{\epsilon_0} = \frac{4}{3} \left(\frac{\mathcal{N}}{2.612 g} \right)^{2/3} \tag{10.61}$$

Similarly the mean occupation of the states with energies $2\epsilon_0$, $3\epsilon_0$, $11\epsilon_0/3$, and $4\epsilon_0$ are $\langle n_0 \rangle/2$, $\langle n_0 \rangle/3$, $3\langle n_0 \rangle/11$, and $\langle n_0 \rangle/4$ respectively. These numbers are all much less than $\mathcal{N}$ for macroscopic systems and this justifies the integral form of (10.56).

We now return to the problem of calculating μ for $T < T_b$. In that temperature range μ has to be positive and less than ϵ_0, so in all the terms in the sum, except the first, we can put $\mu = 0$. The first term refers to the ground state $s = 0$ and if we were to put $\mu = 0$ in that term, it would predict the nonsensical result that the occupation number of the ground state would be negative. This suggests that the conversion of the sum to an integral, while it is justifiable for $s > 0$, does not cope with the occupation of the ground

THE BOSE GAS 159

state. In this term we have to retain μ, and we see that as $T \to 0$ the occupation number of the ground state becomes very large indeed because $\mu \to \epsilon_0$. This then is the phenomenon of the macroscopic occupation of ground state. Thus we write, for $T < T_b$,

$$\mathcal{N} = \langle n_0 \rangle + \frac{2\pi g V}{h^3} (2m)^{3/2} \int_0^\infty \epsilon^{1/2} \left[\exp\left(\frac{\epsilon}{kT}\right) - 1 \right]^{-1} d\epsilon$$

$$= \langle n_0 \rangle + \frac{2\pi g V}{h^3} (2mkT)^{3/2} \int_0^\infty x^{1/2} [\exp(x) - 1]^{-1} dx \quad (10.62)$$

Using (10.58) with $n = \frac{3}{2}$ (remembering that $\zeta(3/2) = 2.612$) and equation (10.59), equation (10.62) can be rearranged to give

$$\langle n_0 \rangle = \left[\exp\left(\frac{\epsilon_0 - \mu}{kT}\right) - 1 \right]^{-1} = \mathcal{N}\left[1 - \left(\frac{T}{T_b}\right)^{3/2} \right] \quad (10.63)$$

This then is the equation that determines μ. It is easily solved to give

$$\mu = \epsilon_0 - kT \ln\left[1 + \frac{1}{\mathcal{N}} \left\{ 1 - \left(\frac{T}{T_b}\right)^{3/2} \right\}^{-1} \right] \quad (10.64)$$

Thus at $T \to 0$, μ differs from ϵ_0 by terms of order $kT/\mathcal{N}$, which are negligible in the thermodynamic limit. However, rather than use μ as the Lagrange multiplier that ensures that we have the correct number of particles, we can alternatively use $\langle n_0 \rangle$. This has the nice physical interpretation that, as we lower the temperature, the number of particles in the ground state increases, to ensure that the total number (strictly speaking the mean total number) of particles remains constant.

In calculating the other thermodynamic properties we must always separate out the ground-state contribution and put $\mu = 0$ in the remaining contributions. Apart from these modifications, the procedure is the same as for the Fermi gas. The results are

$$U = \langle n_0 \rangle \epsilon_0 + \frac{2\pi V g}{h^3} (2m)^{3/2} \int_0^\infty \epsilon^{3/2} \left[\exp\left(\frac{\epsilon}{kT}\right) - 1 \right]^{-1} d\epsilon \quad (10.65)$$

$$= \mathcal{N}\left[1 - \left(\frac{T}{T_b}\right)^{3/2} \right] \epsilon_0 + \frac{3\zeta\left(\frac{5}{2}\right)}{2\zeta\left(\frac{3}{2}\right)} \mathcal{N} kT \left(\frac{T}{T_b}\right)^{3/2}$$

$$= \mathcal{N}\left[1 - \left(\frac{T}{T_b}\right)^{3/2} \right] \epsilon_0 + 0.7701 \mathcal{N} kT \left(\frac{T}{T_b}\right)^{3/2} \quad (10.66)$$

where we have used equations (10.58) and (10.63) and the result that $\zeta(5/2) = 1.34149$. The pressure can be found from the general result (10.35)

$$pV = \frac{2}{3}U = \mathcal{N}\left[1 - \left(\frac{T}{T_b}\right)^{3/2}\right]\frac{2\epsilon_0}{3} + 0.5134\,\mathcal{N}kT\left(\frac{T}{T_b}\right)^{3/2} \tag{10.67}$$

In the thermodynamic limit these equations reduce to

$$U = 0.7701\,\mathcal{N}kT\left(\frac{T}{T_b}\right)^{3/2} \tag{10.68}$$

and

$$pV = 0.5134\,\mathcal{N}kT\left(\frac{T}{T_b}\right)^{3/2} \tag{10.69}$$

The specific heat at constant volume is easily obtained since

$$C_V = \left(\frac{\partial U}{\partial T}\right)_V = 1.931\,\mathcal{N}k\left(\frac{T}{T_b}\right)^{3/2} \tag{10.70}$$

This expression is valid of course for $T < T_b$. For $T > T_b$, it is possible to obtain correction terms to the classical result by expanding in powers of T_b/T (see question 10.2). The result is

$$C_V = \frac{3}{2}\mathcal{N}k\left[1 + 0.231\left(\frac{T_b}{T}\right)^{3/2} + 0.045\left(\frac{T_b}{T}\right)^3 + \ldots\right] \tag{10.71}$$

The Bose gas has a heat capacity which has a cusp at $T = T_b$. This abrupt transition, which is not found for the Fermi gas, is clearly associated with the onset of Bose-Einstein condensation. Figure 10.3 shows C_V over a wide range of temperatures for both Bose and Fermi gases. This and other thermodynamic properties will be discussed when we come to consider real systems in the next chapter.

We have now really calculated all that we need for comparison with experiment. However, we will calculate the entropy because for the Bose gas it illustrates again that care must always be exercised in following an established procedure. For this reason we simply quote the result here, since the detailed derivation is not difficult and is left as an exercise for the reader. The entropy S is

$$k\ln\left[1 - \exp\left(\frac{\mu - \epsilon_0}{kT}\right)\right] + k\left(\frac{\epsilon_0 - \mu}{kT}\right)\left[\exp\left(\frac{\epsilon_0 - \mu}{kT}\right) - 1\right]^{-1}$$

$$+ \frac{5}{2}k\frac{\zeta(5/2)}{\zeta(3/2)}\left(\frac{T}{T_b}\right)^{3/2} \tag{10.72}$$

If we use equation (10.64) to eliminate μ we find that S takes the form

$$k\ln\left\{\mathcal{N}\left[1 - \left(\frac{T}{T_b}\right)^{3/2}\right] + 1\right\} + k\left(\frac{\epsilon_0 - \mu}{kT}\right)\left[1 - \left(\frac{T}{T_b}\right)^{3/2}\right]$$

$$+ \frac{5}{2}k\frac{\zeta(5/2)}{\zeta(3/2)}\left(\frac{T}{T_b}\right)^{3/2} \tag{10.73}$$

Let us consider the limit of this expression as $(T/T_b) \to 0$. The first term approaches $k\ln\mathcal{N}$, the second (making use of the fact that, as we remarked above, ϵ_0 differs from μ by terms of order $kT/\mathcal{N}$) approaches $k/\mathcal{N}$, and the third approaches zero independently of $\mathcal{N}$. The first term therefore dominates so that as $T \to 0$, $S \to k\ln\mathcal{N}$. This is an apparent violation of the third law of thermodynamics, although we should note that the entropy *per particle* vanishes as $k(\ln\mathcal{N})/\mathcal{N}$ in the thermodynamic limit ($\mathcal{N} \to \infty$ while $\mathcal{N}/V$ remains finite). Nevertheless it appears that we do not have a well-defined state at zero temperature, which is curious given that we have made great play about the fact that all the particles condense into the single-particle ground state. We will return to this apparent contradiction in the next section.

We can now construct the Gibbs free energy. This is easily done since $G = U + pV - TS$, we have all the relevant quantities above and we obtain

$$G = kT\ln\left\{\mathcal{N}\left[1 - \left(\frac{T}{T_b}\right)^{3/2}\right] + 1\right\}$$

$$+ \left[(\tfrac{5}{3}\mathcal{N} - 1)\epsilon_0 + \mu\right]\left[1 - \left(\frac{T}{T_b}\right)^{3/2}\right] \tag{10.74}$$

In the thermodynamic limit as $T \to 0$, the term which dominates is

$$G = \tfrac{5}{3}\mathcal{N}\epsilon_0 = \mu\mathcal{N} + \tfrac{2}{3}\mathcal{N}\epsilon_0 \tag{10.75}$$

This then is one of the pathological cases $G \neq \mu\mathcal{N}$ mentioned in Section 4.4. For most systems the second term $\tfrac{2}{3}\mathcal{N}\epsilon_0$ would be negligible since ϵ_0 is proportional to $V^{-2/3}$ but for the Bose gas μ is of the same order and thus

Although we have calculated the entropy, we will do an analogous calculation that led to equation (10.45) for the Fermi gas, since this will give us some physical insight into the behaviour of the Bose gas. Starting from equation (10.13) for the entropy and taking the lower signs, we find

$$S = -k \sum_s [\langle n_s \rangle \ln (\langle n_s \rangle) - (1 + \langle n_s \rangle) \ln (1 + \langle n_s \rangle)] \qquad (10.76)$$

This expression for S, unlike its fermion counterpart, is not transparently interpretable in terms of uncertainties. A little more algebra is necessary as follows. According to the formal treatment of the grand canonical distribution the probability that the system is in a state specified by the set $\{n_s\}$ is

$$P_{\{n_s\}} = \frac{\exp \left[-\sum_s n_s \frac{\epsilon_s - \mu}{kT} \right]}{\Xi} \qquad (10.77)$$

It is worth examining the behaviour of these probabilities in the low-temperature limit. For all states except the ground state the mean occupation numbers go to zero and consequently so do the probabilities that the state will be occupied by any non-zero number of particles. However, for the ground state, the mean occupation number tends to the mean number of particles and consequently for $s = 0$ (the ground state) (10.77) can be written

$$P_{\{n_0\}} = \frac{1}{\mathcal{N} + 1} \left(\frac{\mathcal{N}}{\mathcal{N} + 1} \right)^{\langle n_0 \rangle} \qquad (10.78)$$

For large $\mathcal{N}$ this simplifies to

$$P_{\{n_0\}} = 1/\mathcal{N} \qquad (10.79)$$

The implication of this is that the probability that the ground state will be occupied by a given number of particles is approximately independent of that number. Since all the particles are in the ground state of the system, this means that the probability that the *system* has any number of particles in it is also independent of the number of particles, for example, the probability that the system has 10^{27} particles is practically the same as the probability that it has 10^{31} particles. This equal probability distribution is then the origin of the logarithmic term in the entropy (10.73). It is also the reason for the density fluctuations discussed in the next section.

To return to the formal calculation of the entropy, the uncertainty in the occupancy of the single states is

$$H = - \sum_{n_r=0}^{\infty} P(n_r) \ln P(n_r) \qquad (10.80)$$

On substitution of expression (10.77) we obtain, after rearrangement,

$$H = - \ln \left[\frac{\langle n_r \rangle}{\langle n_r \rangle + 1} \right] \sum_{n_r=0}^{\infty} P(n_r) n_r + \ln (\langle n_r \rangle + 1) \sum_{n_r=0}^{\infty} P(n_r) \qquad (10.81)$$

But $\sum_{n_r=0}^{\infty} P(n_r) n_r = \langle n_r \rangle$ and $\sum_{n_r=0}^{\infty} P(n_r) = 1$, so

$$H = - \langle n_r \rangle \ln \langle n_r \rangle + (\langle n_r \rangle + 1) \ln (\langle n_r \rangle + 1) \qquad (10.82)$$

Summing the uncertainties in occupation for all states and multiplying by k yields equation (10.76) as required. We are now in a position to discuss fluctuations in boson systems.

10.6 Fluctuations in Bose systems

The fluctuations in the occupancy of a single-particle state can be calculated in a similar manner to that for fermions (Section 10.4). The probability that n_s particles occupy single-particle state s is given by equation (10.77). Hence the mean square number of particles in the state s is given by

$$\langle n_s^2 \rangle = \sum_{n_s=0}^{\infty} P(n_s) n_s^2 = \frac{1}{\langle n_s \rangle + 1} \sum_{n_s=0}^{\infty} \left(\frac{\langle n_s \rangle}{\langle n_s \rangle + 1} \right)^{n_s} n_s^2 \qquad (10.83)$$

The sum can be evaluated by setting $x \equiv [\langle n_s \rangle / (\langle n_s \rangle + 1)]$ and using the general result

$$\sum_{n_s=0}^{\infty} n_s^2 x^{n_s} = \frac{x(1-x)}{(1-x)^3}$$

Hence

$$\langle n_s^2 \rangle = \langle n_s \rangle (2 \langle n_s \rangle + 1)$$

and

$$\frac{\langle \Delta n_s \rangle_{\text{rms}}^2}{\langle n_s \rangle^2} = \frac{\langle n_s \rangle + 1}{\langle n_s \rangle} \qquad (10.84)$$

164   GASES IN THE GRAND CANONICAL DISTRIBUTION

For $T \gg T_b$ all $\langle n_s \rangle \to 0$ and hence the fractional deviation is of order ∞. For $T < T_b$, then for $s \neq 0$, $\langle n_s \rangle \sim \mathcal{N}^{2/3}$ and for $s = 0$, $\langle n_s \rangle \sim \mathcal{N}$. Hence, in both cases, the fractional deviation is of order 1. On the other hand, the expression for the fluctuation in the total number of particles is given by arguments similar to that for fermions. The result is

$$\frac{\langle \Delta N \rangle^2_{\text{rms}}}{\mathcal{N}^2} = \frac{1}{\mathcal{N}^2} \sum_s \langle n_s \rangle (\langle n_s \rangle + 1) \tag{10.85}$$

Unlike the Fermi case, this fractional deviation is not small in general, because as $T \to 0$, $\langle n_o \rangle \to \mathcal{N}$ and $\langle n_s \rangle \to 0$ for all $s \neq 0$. Hence

$$\frac{\langle \Delta N \rangle^2_{\text{rms}}}{\mathcal{N}^2} \sim 1 \tag{10.86}$$

This is the result we would have expected from the discussion at the end of the last section, i.e., that the probability that the system has a certain number of particles in it is independent of that number or alternatively that the entropy contains a logarithmic term. Now, according to our discussion in Section 4.3, this would imply macroscopic density fluctuations at temperatures less than T_b. In the next chapter we will consider the application of this model to liquid ^{4}He. For that system such fluctuations are not observed; however in that case there are interactions between the atoms and the occupation numbers of different states are not uncorrelated, i.e., $\langle n_r n_s \rangle \neq \langle n_r \rangle \langle n_s \rangle$. The effect of these correlations will be to reduce the fluctuations in N.

A word of warning is necessary at this point. It would seem reasonable to calculate the fluctuations in N directly from the general expression (4.79) for the fractional deviation. The bulk modulus $B = -V(\partial p/\partial V)_{N,T}$ can be calculated directly from equation (10.69). However, the derivation of equation (4.79) requires a Maxwell relation (see Appendix 2) derived using the result $G = \mu \mathcal{N}$. Now as we have remarked in the previous section, for $T < T_b$, $G \neq \mu \mathcal{N}$ algebraically, so that equation (4.79) is not valid for the perfect Bose gas below T_b.

CHAPTER 10: SUMMARY OF MAIN POINTS

1. It is impossible to evaluate the canonical partition function for either bosons or fermions. However, it is possible to evaluate the grand partition function and this has a different form depending on whether the particles are bosons or fermions.

2. For the classical gas limit, both sets of statistics lead to the classical gas result.

3. For the Fermi gas, the zero temperature value of the chemical potential is known as the Fermi energy. The ground state of that gas is obtained by putting one particle into each of the lowest single-particle states, so that there is only one in each state. The highest energy of the single-particle states that are occupied, is the Fermi energy.
4. At low temperatures the specific heat of the Fermi gas is proportional to the temperature.
5. The Bose gas exhibits a unique phenomenon known as macroscopic occupation of the ground state. Below a certain temperature, known as the Bose temperature, the number of particles in the ground state becomes comparable with the total number of particles. As the temperature approaches zero, the number in the ground state gets closer to the total number of particles.
6. As T approaches zero, the entropy of a Fermi gas approaches zero. For a Bose gas in the same limit the entropy becomes proportional to $\ln \mathcal{N}$ and the gas exhibits macroscopic fluctuations in the total number of particles.

QUESTIONS

10.1 Consider the case of an ideal gas in which the maximum number of particles in any one energy state is p (this is called 'intermediate statistics'). Derive the mean occupation number $\langle n_s \rangle$ in the energy state with energy ϵ_s. Show that the Fermi and Bose formulae follow as special cases.

10.2 Show that at high temperatures the equation of state for a Bose gas is

$$pV = \mathcal{N}kT\left[1 + 0.462\left(\frac{T_b}{T}\right)^{3/2} + 0.023\left(\frac{T_b}{T}\right)^{3} + \ldots\right]$$

and the thermal capacity is

$$C_V = \frac{3}{2}\mathcal{N}k\left[1 + 0.231\left(\frac{T_b}{T}\right)^{3/2} + 0.045\left(\frac{T_b}{T}\right)^{3} + \ldots\right]$$

Obtain similar equations for a Fermi gas.

10.3 Show that, for a Fermi gas, the mean square energy can be written as

$$\frac{4\pi g V}{h^3}(2m)^{3/2}(kT)^{7/2}\int_0^\infty \frac{x^6 dx}{\exp(\alpha + x^2) + 1}$$

where α is a number given by

$$\mathcal{N} = \frac{4\pi g V}{h^3}(2mkT)^{3/2}\int_0^\infty \frac{x^2\,dx}{\exp(\alpha + x^2) + 1}$$

10.4 A crude model for an atomic nucleus is that of a Fermi gas of nucleons in a harmonic oscillator potential (rather than the more usual box). The allowed energies of each nucleon are $\epsilon_r = (r + 3/2)\hbar\omega$ with $r = 0, 1, 2, \ldots$ and the degeneracy of the rth energy level, including factors for spin and isotopic spin, is given by $g_r = 2(r + 2)(r + 1)$.

(a) Find the expression for the density of states $g(\epsilon)$.

(b) Write down an integral expression relating the chemical potential μ to the number $\mathcal{N}$ of nucleons.

(c) Obtain the limiting form, appropriate as $T \to 0$, of your answer to part (b) above. This should be a simple formula relating $\mathcal{N}$ to μ_0.

(d) Sketch a plot of the mean occupation number $\langle n_r \rangle$ versus ϵ in the limit as $T \to 0$.

(e) The energy level spacing $\hbar\omega$ for a given nucleus is 10 MeV (1 MeV $\equiv 1.6 \times 10^{-13}$ J). Its Fermi energy ϵ_f turns out to be 50 MeV. How many nucleons are there in the nucleus?

(f) For $0 < T \ll T_f$ (T_f is the Fermi temperature), the internal energy is

$$U(T) = \mathcal{N}kT_f\left[\frac{3}{5} + \frac{\pi^2}{4}\left(\frac{T}{T_f}\right)^2 \ldots\right]$$

The nucleus of part (d) absorbs 50 MeV from an incident proton. What is its resultant temperature (give kT in MeV)?

(g) What energy must an incident proton give up to the nucleus to transform it to a Boltzmann gas?

10.5 Determine the rate of collisions per unit area of a box containing a gas of fermions at $T = 0$.

10.6 Consider an ideal Bose gas composed of particles which have internal degrees of freedom. It is assumed for simplicity that only the first excited level ϵ_1 of those internal levels has to be taken into account besides the ground-state level $\epsilon_0 = 0$. Determine the Bose–Einstein condensation temperature of this gas as a function of the energy ϵ_1.

10.7 Show that the density of states for a two-dimensional gas is independent of energy and use it to find an exact result for the chemical potential of a two-dimensional Fermi gas.

QUESTIONS   167

10.8 What density of states would be appropriate for dealing with a relativistic Fermi gas, for which $\epsilon^2 = p^2c^2 + m^2c^4$ where c is the speed of light? (See Section 5.5, noting that equation (5.30) remains true while equation (5.31) should be replaced by the relativistic form given above.) For the extreme relativistic case ($\epsilon = pc$) obtain expressions for pV and for U and, without explicitly evaluating them, show that $pV = \frac{1}{3}U$. Finally show that, as $T \to 0$, U approaches the value

$$U = \frac{\pi g c p_0^4}{h^3} V$$

and

$$p_0 = h \left(\frac{3\mathcal{N}}{4\pi g V} \right)^{1/3}$$

10.9 Obtain an expression for the chemical potential of a Fermi gas when $T \ll \mu_0/k$, including the first correction term. See Appendix 5.

10.10 The allowed energies of a particle restricted to two-dimensional motion within a square of side l are

$$\epsilon = \frac{h^2}{8ml^2}(n_x^2 + n_y^2)$$

where n_x and n_y are any non-zero positive integers. Show that the density of states for a two-dimensional gas is independent of energy.

Show, by combining your result with the single-particle occupation numbers for fermions and bosons that the chemical potential for a gas of $\mathcal{N}$ particles is given by

$$\mu = kT \ln \left[\pm \left\{ \exp \left(\pm \frac{T_0 \ln 2}{T} \right) - 1 \right\} \right]$$

where the choice of upper sign ($+$) applies to fermions, the lower ($-$) to bosons, and $T_0 \ln 2 \equiv \mathcal{N} h^2 / [2\pi mkl^2 (2s + 1)]$ defines a convenient characteristic temperature T_0. In deriving the required result, you may quote the standard integrals

$$\int_0^\infty \frac{dx}{b \exp(x) \pm 1} = \pm \ln \left(1 \pm \frac{1}{b} \right)$$

Verify the following table and comment briefly on the similarities and differences between the two-dimensional case discussed here and the three-dimensional case discussed in the text.

	$T = 0$	$T = T_0$	$T \gg T_0$
$\mu_{FD} =$	$+kT_0 \ln 2$	0	$-kT \ln[T/(T_0 \ln 2)]$
$\mu_{BE} =$	0	$-kT_0 \ln 2$	$-kT \ln[T/(T_0 \ln 2)]$

10.11 Consider the degenerate electron 'gas' in a metal as a mixture of two gases, of 'spin-up' and 'spin-down' electrons respectively. When a small field B is applied, a small number of electrons reverse their spins so as to maintain the equality of the chemical potential in the two mixed 'gases'. Show that the resultant net magnetization M is given by

$$M = \frac{3\beta^2 B}{2\epsilon_F} \left(\frac{\mathcal{N}}{V}\right)$$

where β is the magnetic moment of the electron and ϵ_F the chemical potential. You may assume that

$$\epsilon_F = \frac{h^2}{8m} \left(\frac{3\mathcal{N}}{\pi V}\right)^{2/3}$$

and that, within each 'gas', the density of states is

$$g(\epsilon) = \frac{2\pi V}{h^3} (2m)^{1/2} \epsilon^{1/2}$$

REFERENCES

1 For example, Estermann, Simpson and Stern, *Phys. Rev.* (1947), **71**, 238.
2 Leonard, *Phys. Rev.* (1968), **175**, 221.

CHAPTER 11

Real Fermi and Bose Systems

11.1 Introduction
11.2 The 'gas' of conduction electrons in metals
11.3 Liquid helium-4
11.4 Electromagnetic radiation
11.5 Helium-3, the Fermi fluid
Questions
References

11.1 Introduction

It was mentioned in the last chapter that the degenerate regimes $T \ll T_f, T_b$ are inaccessible for almost all real gases because interactions and condensation prevent the valid application of the simple gas model. The rather special cases of spin-polarized monatomic hydrogen, deuterium, and tritium, referred to at the end of Section 9.1, will be mentioned only briefly here with the comment that hydrogen and tritium are bosons, while deuterium is a fermion. In the case of hydrogen, Bose–Einstein condensation can be expected to occur when $T = 1.6(\mathcal{N}/V)^{2/3}$, with T in Kelvin and $(\mathcal{N}/V)$ in atom (nm)$^{-3}$. A realistic example is $(\mathcal{N}/V) = 1 \times 10^{-3}$ atom (nm)$^{-3}$ at 16 mK. Such a number density may not seem high (hydrogen at NTP has $(\mathcal{N}/V) = 27 \times 10^{-3}$ molecule (nm)$^{-3}$) but the experimental difficulties are severe. However, there are several real systems which can be regarded as gaslike in a restricted sense, although they may in fact be solids, liquids or volumes of radiation. We shall see that some of these show strong degeneracy even at temperatures which are not particularly low by room-temperature standards. The condition for strong degeneracy is

$$T \ll T_f \text{ or } T_b, \tag{11.1}$$

where

$$T_f = \frac{h^2}{8mk}\left(\frac{6\mathcal{N}}{\pi V g}\right)^{2/3} \quad \text{(fermions)} \tag{11.2}$$

and

$$T_b = \frac{h^2}{2\pi mk}\left(\frac{\mathcal{N}}{2.612 V g}\right)^{2/3} \quad \text{(bosons)} \tag{11.3}$$

Clearly, regimes of strong degeneracy will be more accessible for systems with low particle mass m and/or high number density $\mathcal{N}/V$ and the following sections of this chapter deal with a number of such cases. In future we may wish to compare results in the grand canonical distribution with those in the canonical distribution. We therefore replace $\mathcal{N}$ by N. In practice there is no difference between the two except at a critical point.

11.2 The 'gas' of conduction electrons in metals

Some properties of metals are explicable in terms of a 'free-electron' model in which the weakly bound valence electrons are treated as a gaseous assembly within the solid lattice. In its simplest version, the electrons are thought of as being free to move within the volume of the lattice and the potential energies of interaction are neglected, except insofar as they determine a collision diameter. As with perfect gases, collisions, though they may be infrequent, play an important role in maintaining thermal equilibrium. This model was used successfully long before the development of quantum mechanics to give a microscopic explanation of Ohm's law and of the Wiedemann–Franz law which expresses the proportionality between electrical and thermal conductivities. However, it was naturally assumed then that the electron gas was describable by Maxwell–Boltzmann statistics and this led to a discrepancy in respect of the heat capacity which, classically, would be expected to add $\frac{3}{2}Nk$ to the lattice contribution discussed in Chapter 6. The experimental absence of the extra term represented a blow against the model, as did the behaviour of the paramagnetic susceptibility of the conduction electrons. Quantum mechanics strengthened the model in two ways: firstly, it led to a better understanding of the motion of electrons in periodic lattices and of metallic bonding, and, secondly, through quantum statistics, it explained the previously baffling heat capacity and paramagnetic susceptibility. These latter properties will be dealt with respectively here and in Section 12.6, while the wider issues are to be found discussed in most textbooks on solid-state physics.

Electrons are fermions with $g = 2$, and in a solid lattice, with at least one free electron per atom, they are densely packed so that T_f (from equation (11.2)) is usually well above the melting point of the metal. The reader can readily check this statement with a few estimates – for example copper, which melts at 1.356×10^3 K, has $T_f = 8.12 \times 10^4$ K, so that its electron gas within the solid is always highly degenerate and we must use quantum statistics to describe its properties. Equations given in Section 10.3 are appropriate, and in particular we shall be interested in their low-temperature limiting forms, equations (10.38) and (10.39):

$$\frac{U}{NkT_f} = \frac{3}{5} + \frac{\pi^2}{4}\left(\frac{T}{T_f}\right)^2 - \ldots \tag{11.4}$$

and

$$C_V = \left(\frac{\partial U}{\partial T}\right)_V = \tfrac{1}{2}\pi^2 Nk \left(\frac{T}{T_f}\right) - \ldots. \tag{11.5}$$

This quantity is generally small compared with the lattice contribution $3Nk$ which is appropriate when $T \gg \theta_D$ (where θ_D is the Debye characteristic temperature equal, for copper, to 343 K). However, when $T \ll \theta_D$, according to the Debye model (Section 6.5),

$$C_V(\text{lattice}) = \frac{12Nk\pi^4}{5}\left(\frac{T}{\theta_D}\right)^3 \tag{11.6}$$

and from equation (11.5) when $T \ll T_f$,

$$C_V(\text{electrons}) = \frac{1}{2}\pi^2 Nk\left(\frac{T}{T_f}\right) \tag{11.7}$$

Consequently, the contributions are equal when

$$T = \frac{\theta_D}{2\pi}\left(\frac{5\theta_D}{6T_f}\right)^{1/2} \tag{11.8}$$

$$\approx 3 \text{ K for copper}$$

Thus, at low temperatures the electronic contribution is indeed important and can be identified by displaying experimental results in a straight-line plot of (C/T) versus T^2 so that the gradient characterizes the lattice and the intercept characterizes the electrons. Figure 11.1 shows such a plot for potassium and illustrates the precision not only of the data but also of the prediction. Incidentally, it is usually the case that the intercept differs in magnitude from the expected value of $\pi^2 Nk/2T_f$ by a small factor of order unity, which is due to the appearance in the solutions of the Schrödinger equation for an electron moving in a periodic lattice, of an 'effective mass'

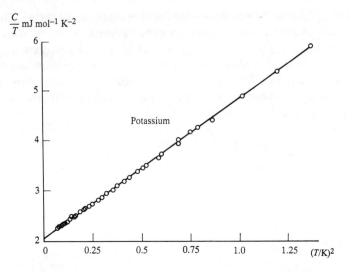

Figure 11.1 The experimental low-temperature heat capacity data for solid potassium obtained and tabulated by W.H. Lien and N.E. Phillips, *Phys. Rev.* (1964), **133**, A1370 who also present data for rubidium and caesium. Data of similar precision for copper, silver and gold is published by W.C. Corak, M.P. Garfunkel, C.B. Satterthwaite and W. Wexler, *Phys. Rev.* (1955), **98**, 1699. The data is plotted as C/T versus T^2 as suggested by equations (11.6) and (11.7), so that the gradient of the plot characterizes the lattice and the intercept the electrons. An insulating crystal such as argon (see Figure 6.2) gives a zero intercept in such a plot.

m^* which differs slightly from the electron 'bare mass' m. This effective mass modifies T_f according to equation (11.2).

We saw in Chapter 10 that, at any temperature, $pV = \frac{2}{3}U$. Consequently, in its limiting form as $T \to 0$, there should be a finite 'gas' pressure given by

$$pV = \frac{2}{5}NkT_f \qquad (11.9)$$

where we have used equation (11.4). The pressure can be unexpectedly large, being about 3.8×10^{10} Pa (i.e., 3.7×10^5 atmospheres) for copper. The strain is taken by the lattice and there is, therefore, no striking evidence of the existence of this pressure, large though it may be. (It should not of course be confused with any pressure measured *outside* the metal, which, whether of electrons or atoms, is tiny – see Section 12.2 and Question 12.6.) However, it is possible in a crude way to regard the observed crystalline spacing as arising from a competition between electrostatic tendency to contract and a Fermi gas tendency to expand. Let us be more specific, remembering that not too much can be expected from so crude a model.

The basic idea is to regard the crystal as a lattice of positive ions, among which the free electrons can wander throughout the whole crystal.

A calculational simplification which does not do violence to the broad idea of the model is to restrict the motion of the electrons to be within an inter-connecting array of touching spheres. If the array were, for instance, hexagonal close-packed, this would reduce the volume available to the electrons to 74 %, but would make it easier to calculate the electrostatic energy because each sphere is neutral and its energy is therefore independent of its neighbours. This is the weakest part of the model because it implies that there is no attraction between neighbouring atoms, which renders the precise meaning of the binding energy calculated below highly problematic. However, with this approach, the calculation of total energy as a function of sphere diameter l is readily performed and we find

$$U_{tot} = U_f + U_{es}, \tag{11.10}$$

where

$$U_f = \frac{3}{5}NkT_f = \frac{3Nh^2}{40ml^2}\left(\frac{18}{\pi^2}\right)^{2/3} \tag{11.11}$$

is the kinetic energy of the electrons from equation (11.4). The electrostatic energy of the array is

$$U_{es} = -\frac{9Ne^2}{20\pi\epsilon_0 l} \tag{11.12}$$

(In this section only, ϵ_0 is the permittivity of free space, not a single-particle ground-state energy.) The total energy is thus attractive at large l and repulsive at small l. The equilibrium spacing l_0 is found by minimizing U_{tot} with respect to l, with the result

$$l_0 = \frac{h^2\pi\epsilon_0}{3me^2}\left(\frac{\pi}{18}\right)^{2/3} = 0.117 \text{ nm} \tag{11.13}$$

This result is rather small by comparison with real metallic lattices (for example, for copper $l_0 = 0.256$ nm), but it is not entirely ridiculous considering the crudity of the model. The minimum value of U_{tot} is given by

$$\frac{U_{tot,0}}{N} = \frac{27me^4}{10\pi^2 h^2 \epsilon_0^2}\left[\frac{81}{\pi^4} - \frac{1}{2}\left(\frac{18}{\pi^2}\right)^{2/3}\right] = 2.54 \text{ eV} \tag{11.14}$$

Again, this is a fairly plausible number (for copper the cohesive energy is experimentally 3.50 eV per atom) although its interpretation is, as mentioned above, somewhat questionable. Anyway, it is clear that the statistics of the Fermi gas are seen to have a strong relevance to the bonding of metallic crystals and to their heat capacities. A further property, that of paramagnetic susceptibility, will be discussed in Section 12.6.

11.3 Liquid helium-4

Figure 11.2 shows the phase diagram of ^{4}He which has at least two remarkable features. Firstly, the liquid phase can exist to the lowest temperatures, a fact due to a combination of two circumstances, the low atomic weight (and consequent high ground-state energy ϵ_0 for an atom 'contained' by its neighbours) and the weak interatomic forces. Secondly, ^{4}He has two quite distinct liquid phases separated by the so-called lambda-line. Below the lambda-line the liquid is referred to as ^{4}He II, above as ^{4}He I; the difference is very marked indeed and the transition is signalled by (among other properties) a heat-capacity anomaly with a logarithmic spike. ^{4}He I is a fairly ordinary, if rather gaslike, liquid; ^{4}He II has very abnormal characteristics including giant heat transport and superfluidity. We shall not attempt here to do more than offer a sketch of the Bose-gas approach to the heat capacity, although a number of other interesting applications of statistical mechanics can be found in the extensive literature of liquid helium.[1]

Let us assume, to begin with, that the Bose-gas model is appropriate to this rarefied liquid, at least as a first approximation. ^{4}He atoms are bosons with $g = 1$, and the expressions for thermodynamic properties are

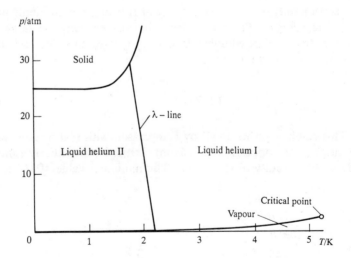

Figure 11.2 The phase diagram of helium-4. This is unusual in two respects. First, that the liquid phase can exist as $T \to 0$ unless considerable pressures are applied, and secondly, that there are two distinguishable liquid phases (He I and He II) separated on the phase diagram by the so-called λ-line.

given in Chapter 10. We must now make specific predictions about the heat capacity. First of all, equation (11.3), with a value of N/V appropriate to liquid ^{4}He, gives

$$T_b = 3.13 \text{ K} \tag{11.15}$$

Actually the vapour-pressure lambda temperature is found experimentally to be 2.17 K which is close enough to be encouraging. According to the model, we must distinguish the regimes $T \leqslant T_b$ and $T \geqslant T_b$ and equations (10.70) and (10.71) give the heat capacity forms for the two regions,

$$C_V = 1.93 Nk \left(\frac{T}{T_b}\right)^{3/2} \tag{11.16}$$

for $T \leqslant T_b$, and

$$C_V = \frac{3}{2} Nk \left[1 + 0.231 \left(\frac{T_b}{T}\right)^{3/2} + 0.045 \left(\frac{T_b}{T}\right)^3 \ldots \right] \tag{11.17}$$

for $T \geqslant T_b$. Figure 11.3 shows the heat capacity plotted against temperature for a perfect Bose gas and for liquid ^{4}He. The comparison cannot be regarded as fully satisfactory. To begin with, the nature of the anomaly is different in the two cases (cusp for the Bose gas, sharp spike for liquid ^{4}He). Also, although according to equation (11.3) T_b rises with density (and hence with pressure), T_λ falls (see Figure 11.2). However we do not wish in these pages to solve *all* difficulties. One difficulty, though, can be explained here with resounding success and it has to do with the heat capacity at temperatures well below the singularity, say below 1.8 K.

According to equation (11.16) the heat capacity should be proportional to $T^{3/2}$. Experimentally it is not, being in general more complicated, but settling down below about 0.6 K to be accurately proportional to T^3. This seems weird because it recalls the behaviour of the Debye *solid* (Section 6.5) and the connection is made stronger by the experimental fact that the magnitude of the heat capacity is found to agree with the Debye expression to within 2 %. This strongly suggests that most of the thermal energy in liquid ^{4}He is carried by vibrational modes (as in solids) and hardly at all by freely-moving gas atoms. This result is confirmed by experiments using neutron scattering techniques which show that the 'excitations' in the liquid have a relationship between energy ϵ and momentum p which is quite different from the particle-type $\epsilon = p^2/2m$. The relationship is shown graphically in Figure 11.4. Let us take stock: we believe we are still dealing with a Bose gas of weakly-interacting excitations below T_b for which the occupation number given by equation (10.11) is still appropriate with the minus sign and with $\mu = 0$. However, according to Section 5.5, the allowed states are specified fundamentally in terms of *momenta*, so the proper density of states should be written in the form

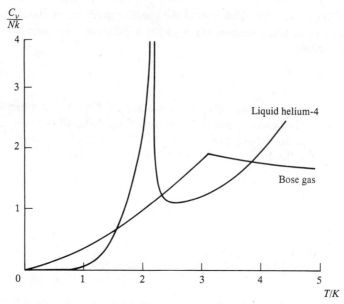

Figure 11.3 Experimental heat capacity of liquid ^{4}He, compared with that of a Bose gas of equal number density using equations (11.16) and (11.17). The liquid data is from a convenient smoothed tabulation given by J. Wilks in Table A1 of *Liquid and Solid Helium* (Oxford, 1967), where the original references are also given.

$$g(p) = \frac{4\pi V}{h^3} p^2 \qquad (11.18)$$

This is functionally similar to $g(\omega)$ for the Debye solid given by equation (6.49). Now our equation for U is given, not by equation (10.66) but by

$$U = \frac{4\pi V}{h^3} \int_0^\infty \epsilon(p) \left\{ \exp\left[\frac{\epsilon(p)}{kT}\right] - 1 \right\}^{-1} p^2 \, dp \qquad (11.19)$$

This may be integrated with $\epsilon(p)$ substituted from the neutron scattering data illustrated in Figure 11.4. More explicitly, it may be noticed that the effect of the bracketed term containing the exponential in the integrand in equation (11.19) is to place disproportionately more weight on the lowest-energy parts of the excitation spectrum which, it turns out, have simple algebraic relations between ϵ and p. Thus, the section near the origin, describing 'phonon' excitations, has $\epsilon = pu$, where u is the velocity of sound in liquid helium; and the section near the minimum, describing 'roton' excitations, has $\epsilon = \Delta + (p - p_0)^2/2\mu$, where Δ, p_0 and μ are known constants characterizing the minimum. Equation (11.19) can now be explicitly integrated to give phonon and roton contributions separately:

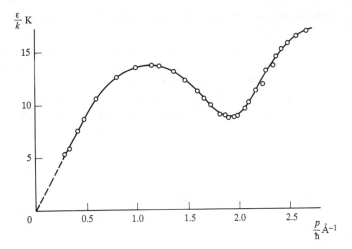

Figure 11.4 The relationship between energy ϵ and momentum p for excitations in liquid ^{4}He II, obtained from neutron scattering experiments and tabulated by D.G. Henshaw and A.D.B. Woods, *Phys. Rev.* (1961), **121** 1266.

$$U = U_{ph} + U_{rot} = \frac{\pi^2 V(kT)^4}{30(\hbar u)^3} + \frac{kTV}{\pi}\left(\frac{1}{2} + \frac{\Delta}{kT}\right)\left(\frac{\mu kT}{2\pi}\right)^{1/2}\frac{p_0}{\hbar^3} \quad (11.20)$$

Hence, by differentiating

$$C_V = C_{V,ph} + C_{V,rot}$$
$$= \frac{2\pi^2 k^4 V}{15\hbar^3 u^3}T^3 + \frac{p_0^2 Vk\Delta}{\pi\hbar^3}\left(\frac{\mu}{2\pi kT}\right)^{1/2}\left[\frac{\Delta}{kT} + 1 + \frac{3kT}{4\Delta}\right]\exp\left(-\frac{\Delta}{kT}\right) \quad (11.21)$$

This expression agrees within a few per cent with experimental results up to about 1.8 K, which is a considerable achievement. Below about 0.6 K, only the first term is important and this is precisely the same result which would have been obtained on applying Debye methods to liquid helium, bearing in mind that a liquid cannot support transverse modes.

The topic of liquid helium is worth inclusion here, not only because of its inherent interest, but also because it raises for the first time a situation in which 'particles' or excitations are *not* described by $\epsilon = p^2/2m$. Another point of general value is the remarkable parallelism of the treatment of the Debye solid and the gas of phonon excitations. In the next section, this will allow a choice of approaches to the question of electromagnetic radiation. Remember that for the Debye solid the canonical distribution was used and each mode was allowed to be in any of the infinite series of harmonic oscillator energy levels. The alternative description is that a mode in the nth level is mathematically the same as n phonons in the first level. Consequently, phonons are to be regarded as bosons (since multiple occupation is allowed) and a boson gas is described here by the grand canonical distribution. Both approaches lead to the same thermodynamic properties. The idea of phonons in solids is well elaborated in standard text books on statistical physics or solid-state physics and is also the topic of Questions 11.10 to 11.12.

11.4 Electromagnetic radiation

There are, as suggested at the end of the last section, two possible ways to proceed.

(a) We may concentrate on the wave-nature of the radiation, arguing that at the boundaries of the containing volume V (at temperature T) certain conditions apply to the solution of the wave equation. For instance, we might consider for convenience a cube of volume $V = L^3$, at the inside surface of which the electric field must be zero as for a metal (the precise boundary conditions considered appropriate make negligible differences to the results for a macroscopic sample but this one is mathematically convenient). The wave equation for E_x, E_y or E_z is formally identical with equation (6.46) relating to displacements in the Debye model. The result for the density of modes function differs from equation (6.49), only because electromagnetic waves have two transverse polarizations whereas the Debye solid may also have a compressional mode. Thus

$$g(\omega) = \frac{V\omega^2}{\pi^2 c^3} \tag{11.22}$$

where c is the velocity of light.

There is no equivalent to equation (6.51) defining the limiting frequency ω_m because in our radiation system there is no requirement that the number of modes be fixed. All frequencies $\omega = 0$ to $\omega = \infty$ are possible. Thus equation (6.40) takes the form

$$\ln Q = -\frac{V}{\pi^2 c^3} \int_0^\infty \omega^2 \ln\left[2\sinh\left(\frac{\hbar\omega}{2kT}\right)\right] d\omega \tag{11.23}$$

There is a mathematical obstacle to overcome in that this integral is divergent. However, it is not difficult to see where this divergence comes from. The logarithmic term in the integrand can be written

ELECTROMAGNETIC RADIATION   179

$$\ln\left[2\sinh\left(\frac{\hbar\omega}{2kT}\right)\right] = \ln\left\{\left[\exp\left(\frac{\hbar\omega}{2kT}\right)\right]\left[1-\exp\left(-\frac{\hbar\omega}{kT}\right)\right]\right\}$$

$$= \frac{\hbar\omega}{2kT} + \ln\left[1-\exp\left(-\frac{\hbar\omega}{kT}\right)\right]$$

The logarithm of the partition function can thus be written

$$\ln Q = -\frac{V}{\pi^2 c^3}\left(\frac{\hbar}{2kT}\right)\int_0^\infty \omega^3 d\omega$$

$$-\frac{V}{\pi^2 c^3}\int_0^\infty \omega^2 \ln\left[1-\exp\left(-\frac{\hbar\omega}{kT}\right)\right]d\omega \qquad (11.24)$$

The second integral coverges but the first diverges. The physical origin of this divergence is clearly the zero-point energy of the waves. For the Debye solid there was a cut-off in the frequency and so the total zero-point energy was finite. For electromagnetic radiation, there exists no cut-off and we have an infinite zero-point energy. This is a well-known anomaly in quantum mechanics, which is dealt with by measuring all energies from the top of the infinite zero-point energy! We should thus simply drop the first term and write

$$\ln Q = -\frac{V}{\pi^2 c^3}\int_0^\infty \omega^2 \ln\left[1-\exp\left(-\frac{\hbar\omega}{kT}\right)\right]d\omega \qquad (11.25)$$

All the thermodynamics can now be obtained in the usual ways appropriate to the canonical distribution. Rather than do that here we pursue the alternative approach.

(b) We concentrate on the particle- (photon-) nature of the radiation, deriving the density of states by converting equation (11.18) by means of the de Broglie relation $p = \hbar/\lambda = \hbar\omega/c$. Hence

$$g(\omega) = \frac{2V\omega^2}{2\pi^2 c^3} \qquad (11.26)$$

The factor 2 again allows for two possible transverse polarizations, and equation (11.26) is identical with equation (11.22). Then we treat the photons as an assembly of bosons in the grand canonical distribution, where significantly the total number of photons is not fixed.

Now in our derivation of the grand canonical distribution, we maximized the uncertainty, subject to the constraint that the mean number of particles is given. If we wish to relax this constraint, then examination of equation (4.80) shows that we must put $\gamma = -\mu/kT = 0$, i.e., for black body radiation, $\mu = 0$. This may be interpreted as meaning that the photon

180   REAL FERMI AND BOSE SYSTEMS

Bose gas is always below T_b, that is, it is degenerate at all temperatures. Thus,

$$\ln \Xi = -\int_0^\infty g(\omega) \ln\left[1 - \exp\left(-\frac{\hbar\omega}{kT}\right)\right] d\omega$$

$$= -\frac{V}{\pi^2 c^3} \int_0^\infty \omega^2 \ln\left[1 - \exp\left(-\frac{\hbar\omega}{kT}\right)\right] d\omega \qquad (11.27)$$

This can be integrated by parts to give

$$\ln \Xi = \frac{8\pi^5 V}{45}\left(\frac{kT}{hc}\right)^3 \qquad (11.28)$$

Hence, by appropriate operations on $\ln \Xi$ we find

$$pV = \frac{1}{3} U = \frac{8\pi^5 V(kT)^4}{45(hc)^3} \qquad (11.29)$$

and

$$C_V = \frac{32\pi^5 Vk}{15}\left(\frac{kT}{hc}\right)^3 \qquad (11.30)$$

Note that $pV = \frac{1}{3}U$ in contrast with the perfect gas relation $pV = \frac{2}{3}U$. This has to do with the relativistic character of photons, i.e., the energy is proportional to the momentum rather than the momentum squared, and remains in any case in agreement with the form $pV = \alpha U$ derived in Chapter 2 on more general grounds. The heat capacity is proportional to T^3 at *all* temperatures, which is not the case for Debye solids which have an upper limiting frequency arising from their lattice periodicity. The radiation pressure given by equation (11.29) is generally not large in magnitude except at extreme temperatures (say, 10^5 K or greater) and is independent of the volume.

We shall complete this discussion of radiation by deriving the Planck, Wien and Stefan–Boltzmann radiation laws. The Planck law describes the distributions of energy among the infinite range of possible frequencies and is easily derived by writing the internal energy U as an integral over ω. Thus, using equation (10.65) in the thermodynamic limit,

$$U = \int_0^\infty u(\omega)\, d\omega \qquad (11.31)$$

where

$$u(\omega) = \frac{V\hbar}{\pi^2 c^3} \frac{\omega^3}{\exp\left(\frac{\hbar\omega}{kT}\right) - 1} \qquad (11.32)$$

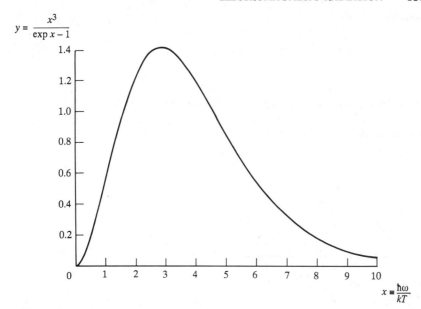

Figure 11.5 The form of the radiation distribution function $u(\omega)$ defined in equations (11.31) and (11.32) can be conveniently displayed in dimensionless form by plotting

$$y(x) \equiv \frac{x^3}{\exp(x) - 1} \quad \text{where} \quad x \equiv \frac{\hbar\omega}{kT}$$

With this notation, equation (11.32) takes the form

$$u(\omega) = \hbar\pi V \left(\frac{kT}{\hbar\pi c}\right)^3 y(x)$$

Equation (11.32) agrees with experiment and its historical importance is that it was first obtained empirically in a celebrated paper of 1900 by Planck long before the arrival of quantum mechanics. It was already known that the high-temperature limiting form of (11.32), $V\omega^2 kT/\pi^2 c^3$, could be derived by classical methods and Planck was able to explain the breakdown of this form only by postulating the quantization of the harmonic oscillator energies. In 1900, this was a most remarkable step to take.

Expression (11.32) is plotted in a convenient dimensionless form in Figure 11.5 and it is seen to have a maximum value. From equation (11.32), this maximum value is found by differentiating with respect to ω and equating the derivative to zero. Solving for ω gives the value of ω for which u is a maximum.

$$u(\omega_m) = 1.421 \frac{V\hbar}{\pi^2 c^3} \left(\frac{kT}{\hbar}\right)^3 \tag{11.33}$$

where

$$\omega_m = 3.69 \times 10^{11} \, T \text{ radians per second.} \qquad (11.34)$$

Equations (11.33) and (11.34) are known as 'Wien's laws', which are often expressed alternatively in terms of the wavelength $\lambda = 2\pi c/\omega$.

Finally, we know that $\mu = 0$, so we can find the expectation number of photons from equation (10.62) in the thermodynamic limit:

$$\langle N \rangle = \int_0^\infty n(\omega) \, d\omega \qquad (11.35)$$

where

$$n(\omega) = \frac{V\omega^2}{\pi^2 c^3} \frac{1}{\exp\left(\dfrac{\hbar\omega}{kT}\right) - 1} \qquad (11.36)$$

The expectation value of the energy per photon of frequency ω is thus

$$\langle \epsilon \rangle = \frac{u(\omega)}{n(\omega)} = \hbar\omega \qquad (11.37)$$

Now, to derive the Stefan–Boltzmann law, we take as our model of a black body a small hole of area A in the vessel containing the radiation. According to familiar kinetic theory arguments (see Question 11.1) the rate of escape of photons with angular frequencies in the range $d\omega$ at ω is $cAn(\omega)d\omega/4V$. Each photon has an energy $\hbar\omega$ and hence the total power radiated per unit area is

$$P = \tfrac{1}{4}c \int_0^\infty \frac{n(\omega)}{V} \hbar\omega \, d\omega \qquad (11.38)$$

$$= \tfrac{1}{4}c \int_0^\infty \frac{u(\omega)}{V} \, d\omega \qquad (11.39)$$

$$= \tfrac{1}{4}c \frac{U}{V} \qquad (11.40)$$

$$= \frac{\pi^2 k^4}{60 \hbar^3 c^2} T^4 = \sigma T^4 \qquad (11.41)$$

using equation (11.29). This is the Stefan–Boltzmann radiation law and σ is the Stefan constant, whose value is $5.65 \times 10^{-8} \, \text{W m}^{-2} \, \text{K}^{-4}$.

It is worth remarking here that the discussion of fractional deviations in a boson gas given in Section 10.6 ought to be applicable to the photon gas. (It also might have been expected to apply to liquid ^{4}He, but in fact interatomic interactions obscure the effect in that case.) Experiments[2] have been done using photoelectric detectors sensitive to the arrivals of single photons, and it has been shown[3] that the fluctuations of the photon counts recorded by a photoelectric detector which is illuminated by light from a black body are just those to be expected from a boson assembly. Alternative descriptions in terms of the wave model are also possible.

Finally, it has been hinted in Section 6.6 and above that an alternative description of lattice modes in a solid is to treat them as a gas of phonons (the analogy with photons is close, particularly at low temperatures). This approach is the topic of Questions 11.10 to 11.12.

11.5 Helium-3, the Fermi fluid

^{3}He atoms are fermions with $g = 2$. The fluid has a critical temperature of 3.3 K and, like liquid ^{4}He, remains liquid as $T \to 0$ unless considerable pressures are applied. A comparison of its properties with those of the perfect Fermi gas is, at first, disappointing, since, although T_f is estimated to be 5.0 K, true degenerate Fermi behaviour is not shown until the temperature is reduced below about 0.05 K. This is reasonably well understood in terms of interactions and 'quasi-particles' – there are no phonons or rotons – but its elucidation involves more argument than we wish to include here. Suffice it to say that, when $T \leqslant 0.05$ K, the heat capacity is closely proportional to T and that the temperature dependence of other properties is broadly in line with Fermi gas predictions.

There exist much better ^{3}He systems showing Fermi characteristics, namely dilute solutions of ^{3}He in liquid ^{4}He. Experiment shows that, provided there is less than about 6 % ^{3}He in the mixture, it can be cooled to any temperature without phase-separation occurring. The whole phase-separation diagram is shown in Figure 12.3 in the next chapter, where we shall apply statistical mechanics to parts of it. In this section, we concentrate on the left-hand corner describing low-temperature dilute solutions. The beauty of such systems arises from two simple facts: firstly, that the solute ^{3}He atoms behave like particles to a very good approximation and secondly, the solvent liquid ^{4}He has a heat capacity which falls rapidly with temperature (equation (11.21) and Figure 11.3) to very tiny values, relative to those of realistic concentrations of ^{3}He. Furthermore, the liquid ^{4}He has zero viscosity. The consequence of these facts is that the solute ^{3}He behaves like a Fermi *gas* whose heat capacity dominates the heat capacity of the solution even for concentrations as small as 0.1 % ^{3}He. Thus for the first time there is available a system which in some respects is just like a perfect Fermi gas *which does not condense as* $T \to 0$ and whose Fermi temperature T_f can be adjusted within limits by varying the ³He percentage concentration $X = (100N_3/(N_3 + N_4))$. A substitution of appropriate numerical values in equation (11.2) gives

$$T_f = 117 X^{2/3} \text{ mK} \tag{11.42}$$

Now since, experimentally, temperatures as low as 10 mK are readily available, a mixture with $X = 1\%$ ($T_f = 117$ mK) can be cooled into the fully-degenerate temperature region or warmed to the classical region at 1 K or above. The heat capacity would thus be expected to be given by equation (11.5) at low temperatures ($T \ll T_f$) and by $\frac{3}{2}N_3 k$ at high temperatures ($T \gg T_f$). To cover the whole range we should use equations (10.34) and (10.35):

$$pV = \tfrac{2}{3}U = \frac{8\pi V kT}{3\hbar^3}(2m)^{\frac{3}{2}} \int_0^\infty \epsilon^{\frac{3}{2}} \left\{ \exp\left[\frac{\epsilon - \mu}{kT}\right] + 1 \right\}^{-1} d\epsilon \tag{11.43}$$

where μ is determined from

$$N_3 = \frac{4\pi V}{\hbar^3}(2m)^{\frac{3}{2}} \int_0^\infty \epsilon^{\frac{1}{2}} \left\{ \exp\left[\frac{\epsilon - \mu}{kT}\right] + 1 \right\}^{-1} d\epsilon \tag{11.44}$$

U is tabulated in Appendix 4 and its gradient with respect to temperature is equal to the heat capacity. Such tables[4] have been available since 1939, but it was not until 1966 that any system was found experimentally (see Figure 11.6) to which the tables were applicable except in the low temperature (conduction electrons) or high temperature (classical gas) regimes. Figure 11.6 shows the success of the result. Equation (11.43) also yields an equation of state in terms of pressure p based, of course, on the perfect gas model. However, liquid solutions are not gases so the meaning of p is not immediately clear and it must certainly not be confused with the vapour pressure. With dilute ³He solutions it can in fact be identified as the osmotic pressure Π and measured by using a very fine channel (superleak) as the semipermeable membrane. This freely allows the flow of the ⁴He solvent (with zero viscosity) but effectively blocks the passage of ³He. Figure 11.7 shows how the osmotic pressure Π might be measured. The low-temperature limiting forms of equation (11.43) are

$$\Pi V = N_3 kT \quad \text{when} \quad T \gg T_f \tag{11.45}$$

which implies that $\Pi \propto XT$ at high temperatures, and

$$\Pi V = \tfrac{2}{5}N_3 kT_f \quad \text{when} \quad T \ll T_f \tag{11.46}$$

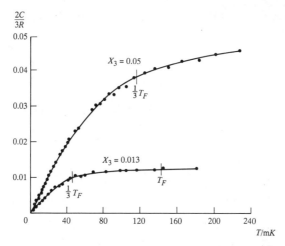

Figure 11.6 Specific heat of mixtures below 230 mK, plotted as $\frac{2C}{3R}$. Two concentrations X are represented. The changeover from classical perfect gas-like behaviour ($\frac{2C}{3R} = X$) at higher T to the linear behaviour of degenerate Fermi gas-like behaviour ($\frac{2C}{3R} = \frac{1}{3}\pi^2 X(T/T_f)$) at lower T is seen to occur in the region of $T = \frac{1}{3}T_f$. (Ref: A.C. Anderson, D.O. Edwards, W.R. Roach, R.E. Sarwinski, and J.C. Wheatley, *Phys. Rev. Lett.* (1966), **17**, 367.)

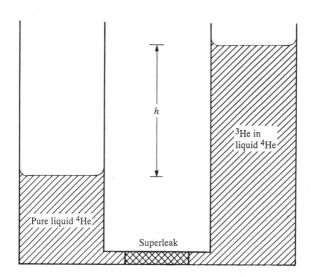

Figure 11.7 A notional arrangement for measuring the osmotic pressure of dilute ^{3}He solution in liquid ^{4}He II relative to pure liquid ^{4}He II. The superleak will pass liquid ^{4}He II readily because of its superfluidity but will hold back the ^{3}He for a period well in excess of the time required to make a measurement. In practice, the height h is likely to be so large that other methods of measuring pressure have to be used.

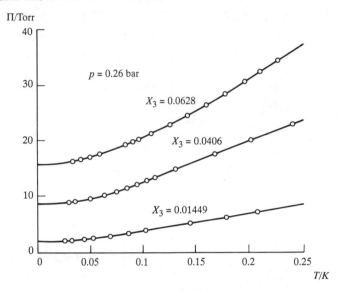

Figure 11.8 Osmotic pressures of some dilute mixtures at a pressure of 0.26 bar. Note that the pressure axis is labelled in torr, so the pressures are quite large. (Ref: J. Landau, J.T. Tough, N.R. Brubaker, and D.O. Edwards, *Phys. Rev.* (1970), **2A**, 2472.)

which implies that $\Pi \propto X^{5/3}$ independently of T at low temperatures.

Equation (11.45), called van't Hoff's law which we shall re-derive in Chapter 12, has been found to be accurate to within a few per cent for concentrations between $X = 0.135\%$ and $X = 4.12\%$ at temperatures between 1.22 K and 1.9 K. Perhaps more interestingly, an examination of the region $T \lesssim T_f$ has shown (see Figure 11.8) that, qualitatively at least, equation (11.43) is right and the limiting form (11.46) is approached at low enough temperatures. Furthermore, the dependence of osmotic pressure on X and T is broadly correct although there are quantitative differences which tend to disappear at the lowest concentrations and to rise to about 50% at $X = 6\%$. The origin of the discrepancies is related to the interactions between ^{3}He atoms in solution which have been totally ignored in this account, and a study of the discrepancies in this and other properties has led gradually to a better understanding of these interactions.

This chapter has attempted to relate some of the observed thermodynamic properties of systems as diverse as metals, liquid helium and radiation to the predictions made for perfect gases in the grand canonical distribution. The main part of this book is now complete, and the three final chapters are intended to serve as introductions to three most important areas which cannot be fully covered in a work of this length, and which in any case are commonly omitted in undergraduate courses. The first topic is that of two-phase assemblies, the second that of interacting gases and real equations of state such as the van der Waals equation and the third, phase transitions.

CHAPTER 11: SUMMARY OF MAIN POINTS

1. This chapter is devoted to an exploration of a wide range of phenomena which can be understood only in terms of Fermi–Dirac or Bose–Einstein statistics with low-temperature degeneracy as a significant effect.
2. There are characteristic degeneracy temperatures T_f for fermions and T_b for bosons, which can be calculated from equations (11.2) and (11.3) respectively in terms of the particle mass and number density.
3. An important example is that of the gas-like behaviour of free electrons in metals; free electrons have a low mass and a high number density and the degeneracy temperature can be tens of thousands of degrees above room temperature. Expressions for the internal energy, specific heat, and gas pressure are obtained, and the jellium model of crystal binding in metals is described.
4. The dramatic properties of liquid ^{4}He are compared with those of a degenerate gas of bosons, and the liquid does exhibit a phase transition which is related to Bose–Einstein condensation.
5. The model is taken further by allowing for the fact that liquid ^{4}He is better described in terms of excitations (phonons and rotons) rather than atoms, and this improves the detailed agreement considerably.
6. A gas of photons can be treated as a degenerate gas of massless bosons and it is shown how this leads to the known radiation laws of Planck, Wien, and Stefan–Boltzmann.
7. Finally, the lighter isotope ^{3}He of helium allows us to compare the degenerate Fermi–Dirac gas with the real liquid ^{3}He. This can be either pure liquid ^{3}He or dilute solutions of ^{3}He in liquid ^{4}He. The latter case in particular, in which the degeneracy temperature can be set by a choice of the concentration of the mixture, provides very striking confirmation of the smooth transition from degenerate to classical behaviour; for example, a mixture containing 1 % of ^{3}He has $T_f = 117$ mK and modern experimental techniques allow experiments over a range from a few mK to a few K.

QUESTIONS

11.1 Show that the number of photons with angular frequencies in the range $d\omega$ at ω arriving from all possible directions at an area A on the inside wall of a box containing radiation is $\frac{1}{4}\rho(\omega)c\,d\omega$ where $\rho(\omega)$ is the number density ($\rho(\omega) \equiv n(\omega)/V$) of such photons in the box.

11.2 Give numerical estimates for the Fermi energy of

188   REAL FERMI AND BOSE SYSTEMS

    (a)   electrons in a typical metal,

    (b)   nucleons in a heavy nucleus,

    (c) ^{3}He atoms in liquid ^{3}He (atomic volume = 0.0462 nm^3 per atom).

Treat all the mentioned particles as free particles.

11.3 Obtain an expression for the bulk modulus associated with the gas of conduction electrons, and make a numerical comparison with the experimental bulk modulus (3.33×10^9 N m^{-2}) of metallic potassium which has a free electron concentration of 1.40×10^{28} m^{-3}.

11.4 Aluminium has an electron concentration of 1.81×10^{29} m^{-3}. Show that electrons of energy equal to the Fermi energy move with speed 2.01×10^6 m s^{-1}.

    (a)   To what extent is the mass of such an electron relativistically increased?

    (b)   Compare its de Broglie wavelength with the lattice spacing of aluminium (0.405 nm). Does the comparison shake your confidence in the validity of the free electron model which treats electrons as particles?

11.5 Use equation (A5.13) in Appendix 5 to estimate the slight downward movement of the Fermi level (i.e., chemical potential μ) as metallic sodium (electron concentration 2.65×10^{28} m^{-3}) is warmed from $T = 0$ K to $T = 300$ K. Does your result justify the common use of Fermi energy as a characteristic constant for a given metal?

11.6 Aluminium has atomic weight 27, valency 3, density 2.7×10^3 kg m^{-3} and a Debye characteristic temperature $\theta_D = 385$ K. Show that the molar heat capacity includes a contribution of roughly 0.4 % from the free electron gas at 300 K, using the tabulated results from Question 6.13 and equation (10.39).

11.7 Determine the following quantities for iron:

    (a)   The probability that a single-particle translational energy state is occupied by a valence electron if the energy state lies 0.1 eV and 1.0 eV above the Fermi energy of 11.5 eV when the iron is at a temperature of 300 K.

    (b)   The temperature to which the iron must be raised in order that the probability of occupancy by a valence electron of a single-particle translational energy state that lies 0.5 eV above the Fermi energy level is 0.01.

(c) The separation in energy between two single-particle translational energy states whose probabilities of occupancy by a valence electron are 0.1 and 0.9 respectively, expressing your answer in units of kT.

11.8 Assuming the sun to radiate as a black body, calculate

(a) the total emission from the surface of the sun ($R = 7 \times 10^5$ km, $T = 6000$K),

(b) the predominant frequency ω_m,

(c) the radiation pressure in the interior ($T = 10^7$ K).

11.9 (a) The sun contains 1400 kg m^{-3} of hydrogen at an interior temperature of about 10^7 K. Prove (see Section 9.5 and Question 9.16) that the hydrogen is fully ionized and that it is a reasonable approximation to treat the electron and proton gas as classical gases.

(b) Estimate the relative importance of electron gas pressure, proton gas pressure and radiation pressure in opposing gravitational collapse.

(c) Suppose the sun somehow managed to collapse to become a pulsar of radius 10 km, the interior temperature remaining the same. Show that the electron gas would be Fermi degenerate *and* extremely relativistic (i.e., $\epsilon = pc$) and show that the Fermi energy would be about 10^8 eV (see Question 10.8).

(d) Given that the energy release in the decay reaction $n \rightarrow (p + e^-)$ is only 8×10^5 eV, explain in broad terms why pulsars are believed to be largely composed of neutrons rather than a mixture of electrons and protons, as assumed in part (c).

(e) Set up a detailed analysis of the gas reaction $n \rightarrow (p + e^-)$ along the lines of Section 9.5 but applicable to a situation where neutrons, protons and electrons are Fermi degenerate and extremely relativistic (i.e., $\epsilon = pc$).

11.10 The approach in Chapter 6 to the Einstein, Debye and lattice-dynamic models of solids was based on the results of the canonical distribution. However, as we have said elsewhere, one is generally free to choose one's formulation on the basis of mathematical convenience alone, and we could have chosen to apply the grand canonical distribution to any of the models. The truth of this statement is illustrated in Chapter 7 where two different approaches were compared for a paramagnetic solid. Again in Chapter 11, we have applied two methods to the box-of-radiation. You should now attempt to deal with the Debye solid, regarding it as a gas of *phonons* with $\epsilon = h\nu$ where $\nu\lambda = c$, the

REAL FERMI AND BOSE SYSTEMS

velocity of sound. As with photons, the number is not fixed but adjusts to satisfy $(\partial F/\partial N)_{V,T} = 0$, i.e., $\mu = 0$, and also since a mode with energy $nh\nu$ is regarded as being equivalently described as n phonons of energy $h\nu$, there is no limitation on occupation. Consequently you should treat the Debye solid as a gas of bosons below $T = T_b$, obtaining an expression for heat capacity. Why does the T^3 dependence break down at high temperatures, which is not the case for photons?

11.11 Aluminium has atomic weight 27, density $2.7 \times 10^3 \text{ kg m}^{-3}$ and an elastic wave velocity (much the same for transverse and longitudinal modes) of $3.3 \times 10^3 \text{ m s}^{-1}$. Using these data, calculate the Debye temperature θ_D and the phonon gas pressure within aluminium at 96.5 K. The idea of phonons is described in the previous question.

11.12 Some experimental values for the heat capacity of liquid ^{4}He are given in the table below and may be taken as isochoric results.

Temperature in K	Heat capacity in mJ mol^{-1} deg^{-1}
0.20	0.65
0.25	1.28
0.30	2.20
0.35	3.50
0.40	5.22
0.45	7.44
0.50	10.20
0.55	13.58
0.60	17.63

(a) Show that the behaviour of the heat capacity is characteristic of that of a gas of phonons.

(b) Find the velocity of sound in liquid ^{4}He below 0.6 K.

11.13 A gas of ^{3}He atoms at 300 K and 1 bar pressure behaves almost perfectly as a classical gas. On the other hand *liquid* ^{3}He at 0.1 K behaves very much like a degenerate Fermi gas. Consider the following idealization. A gas of ^{3}He atoms is held in a closed volume V with a fixed density 83 kg m^{-3}. It can be maintained at any chosen temperature between 300 K and $T \to 0$ K and is assumed always to behave perfectly. Calculate

(a) its chemical potential (in joules) as $T \to 0$ K,
(b) the temperature at which its chemical potential is zero,
(c) its chemical potential at $T = 300$ K, and
(d) its pressure (in pascals) in each case.

REFERENCES

1. For example, Wilks and Betts, *An Introduction to Liquid Helium*, 2nd ed. (Oxford, 1987).
2. Hanbury Brown and Twiss, *Nature* (1956), **177**, 27.
3. Kahn, *Optica Acta* (1958), **5**, 93.
4. Stoner E.C., *Phil. Mag.* (1939), **28**, 257.

CHAPTER 12
Equilibrium in Two-phase Assemblies

12.1	The chemical potential	12.7	Contact potential
12.2	Vapour pressure of an Einstein solid	12.8	Surfaces
		12.9	Adsorption isotherms
12.3	The ideal mixture	12.10	Two-dimensional solid layers
12.4	Phase separation of mixtures	12.11	Two-dimensional gaseous layers
12.5	Osmotic pressure		
12.6	Electron paramagnetism		Questions

12.1 The chemical potential

We have already referred to the chemical potential in several places, most importantly as an identifiable parameter in the grand canonical distribution. The identification of the Lagrange multiplier γ as $-\mu/kT$ in Section 4.4 arose from the comparison of a statistical mechanical expression, equation (4.64), with a quoted thermodynamic relation. We now wish to be more explicit about the thermodynamic properties of the chemical potential μ.

We can write the first and second laws of thermodynamics, for the common but restricted case in which N is fixed, as

$$dU = TdS - pdV \qquad (12.1)$$

The case of N fixed is suitable for single-phase systems, but for a two-phase system of, say, liquid and vapour, we may wish to know whether the total system would reduce or increase its energy by evaporation of liquid. This is a situation in which we would need to compare the loss of energy by the liquid associated with the departure of a number of atoms, with the gain in energy by the vapour associated with the arrival of the same number. For this more general situation, equation (12.1) needs modifying by the addition of an extra term:

$$dU = TdS - pdV + \mu dN \tag{12.2}$$

Thus

$$\mu = \left(\frac{\partial U}{\partial N}\right)_{S,V} \tag{12.3}$$

is defined as the change in internal energy when one particle is added at constant volume and entropy. The requirement of constant entropy S makes this an inconvenient expression for many purposes and an equivalent form can be found in terms of the Helmholtz free energy defined by

$$F = U - TS \tag{12.4}$$

Differentiating,

$$dF = dU - TdS - SdT \tag{12.5}$$

$$= -pdV + \mu dN - SdT \tag{12.6}$$

using equation (12.2). Therefore

$$\mu = \left(\frac{\partial F}{\partial N}\right)_{V,T} \tag{12.7}$$

Thus, using the canonical bridge equation $F = -kT \ln Q$,

$$\mu = -kT \left(\frac{\partial \ln Q}{\partial N}\right)_{V,T} \tag{12.8}$$

Equation (12.7) states that μ is the increase in Helmholtz free energy of a system on the addition of one particle under the conditions of constant volume and temperature. Alternative and equivalent descriptions can also be given in terms of internal energy (equation (12.3)) or Gibbs free energy (see below, equation (12.12)). Equation (12.8) enables us to obtain expressions for μ in the canonical distribution where Q is known. There is a particularly close relationship between μ and the Gibbs free energy G defined by

$$G \equiv U + pV - TS \tag{12.9}$$

Differentiating,

$$dG = dU + p\,dV + V\,dp - T\,dS - S\,dT$$

Hence, using equation (12.2)

$$dG = \mu\,dN + V\,dp - S\,dT \tag{12.10}$$

Thus, for fixed N, p, and T, the Gibbs free energy is fixed. Now in most cases of macroscopic systems G must be extensive (that is, proportional to N). The reader should remember that there are pathological exceptions to this rule, and we have referred to these in Sections 4.4, 7.3, and 10.5. Since from equation (12.10) G is a function of N, p, and T, it follows that

$$G = N\varphi(p, T) \tag{12.11}$$

where $\varphi(p, T)$ is a property of the system which depends on the intensive quantities p and T. From equations (12.20) and (12.11),

$$\mu = \left(\frac{\partial G}{\partial N}\right)_{p,T} = \varphi(p, T) = \frac{G}{N} \tag{12.12}$$

Thus, μ is the increase in Gibbs free energy of the system on addition of one particle under the conditions of constant pressure and temperature or, better still, it is the *Gibbs free energy per particle of the system*. The equivalent italic statements could not be made about the internal energy

$$U = N\varphi_1\left(\frac{S}{N}, \frac{V}{N}\right)$$

or Helmholtz free energy

$$F = N\varphi_2\left(\frac{V}{N}, T\right)$$

because of the appearance of N inside the functional brackets (see Question 12.1). Finally the combination of equation (12.12) and the differential of (12.10) gives a further useful result, the Gibbs–Duhem relation,

$$N\,d\mu = V\,dp - S\,dT \tag{12.13}$$

The conditions of equilibrium in a single phase were derived from statistical mechanical reasoning in Sections 4.3 and 4.4 for the canonical and

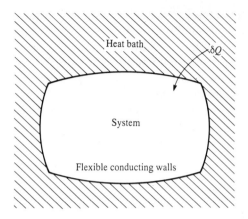

Figure 12.1 A system with flexible, heat-transmitting walls in an infinitely large heat bath.

grand canonical distributions and in Question 4.2 for the 'constant pressure' or petit canonical distribution. In the latter case, it is found that the equilibrium situation minimizes the Gibbs free energy and this may also be derived from purely thermodynamic reasoning. Referring to Figure 12.1 showing a flexible-walled system in a large heat bath, we now imagine a process whereby an amount of heat δQ passes from the bath to the system. Treating the heat bath as an infinite reservoir, we conclude that its loss of δQ is reversible so that its entropy change is $-\delta Q/T$. However, for the finite system the process is not reversible and we therefore denote *its* entropy change by δS. Then, by the second law,

$$\delta S - \frac{\delta Q}{T} \geq 0 \tag{12.14}$$

But for the system, from the first law,

$$\delta Q = \delta U + p\delta V \tag{12.15}$$

Combining these gives

$$T\delta S - \delta U - p\delta V \geq 0 \tag{12.16}$$

We can readily treat four cases. First, for a process in which S and V are constrained to remain constant, equation (12.16) reduces to

$$\delta U \leq 0 \tag{12.17}$$

which implies that a single-phase system subject to these constraints progresses towards equilibrium by reducing the internal energy U. Equilibrium is thus characterized in this case by a minimum value of U. This is not a common situation in practice because of the specified constraint on S, but it does arise in the consideration of the equilibrium state of two vessels filled with superfluid helium linked by a 'superleak' which to a good approximation blocks the flow of entropy from one vessel to the other.

The second and third cases are far more common in practice. For a process in which T and V are constrained to remain constant, equation (12.16) reduces to

$$\delta(U - TS) \leqslant 0$$

that is,

$$\delta F \leqslant 0 \tag{12.18}$$

which implies that a single-phase system subject to these easily applied constraints moves towards equilibrium by reducing the Helmholtz free energy F. Equilibrium is thus characterized in this case by a minimum value of F. Finally, for a process in which T and p are constrained to remain constant (a chemical reaction in an open test tube, for example), equation (12.16) reduces to

$$\delta(U + pV - TS) \leqslant 0$$

that is,

$$\delta G \leqslant 0 \tag{12.19}$$

which implies that a single-phase system, subject to these common constraints, moves towards equilibrium by reducing the Gibbs free energy G. Equilibrium is thus characterized in this case by a minimum value of G. The fourth case can similarly be constructed, in which the enthalpy $H = U + pV$ is minimized if p and S are fixed. That is,

$$\delta H \leqslant 0 \tag{12.20}$$

For two phases 1 and 2 (for example, liquid and solid) the condition of equilibrium is the same, namely $\mu_1 = \mu_2$, whichever of the specified four sets of constraints are applied. The proof of this is as follows. Clearly a two-phase system, like a single-phase system, moves towards equilibrium by reducing the appropriate energy, U, F, G, or H, depending on the constraints. Now the two-phase (and by extension the multiphase) system may adopt a strategy not available to the single-phase system for free energy reduction. This strategy is the transfer of atoms or molecules from one phase to the other. When equilibrium is reached it must be the case that transfer of a small number of atoms or molecules would result in, say, an increase in the appropriate energy in one phase (say 1) which is exactly balanced by a reduction in that energy in the other (say 2); the net change must be zero. Let the appropriate energy ($U, F, G,$ or H) be denoted by X. So if, at equilibrium, a few atoms or molecules change sides, the net change in energy must be zero:

$$\left(\frac{\partial X}{\partial N_1}\right)_{\text{(constraints)}} dN_1 + \left(\frac{\partial X}{\partial N_2}\right)_{\text{(constraints)}} dN_2 = 0 \tag{12.21}$$

But also, because atoms or molecules are conserved,

$$N_1 + N_2 = \text{constant} \tag{12.22}$$

so that

$$dN_1 = -dN_2 \tag{12.23}$$

and so, from (12.21),

$$\left(\frac{\partial X}{\partial N_1}\right)_{\text{(constraints)}} - \left(\frac{\partial X}{\partial N_2}\right)_{\text{(constraints)}} = 0 \tag{12.24}$$

Now, from this equation and taking the appropriate energies $X = U, F, G,$ or H, together with equations (12.3), (12.7), (12.12), and a similar one involving H, we deduce that $\mu_1 - \mu_2 = 0$, i.e.,

$$\mu_1 = \mu_2 \tag{12.25}$$

This is the condition of equilibrium between phases. It proves to offer a powerful means of calculating physical parameters such as vapour pressure or magnetic susceptibility when two or more phases are in mutual equilibrium. The remainder of this chapter is devoted to examples of this.

12.2 Vapour pressure of an Einstein solid

The choice of an Einstein solid is just to simplify the algebra, but the reader can easily extend the approach to any other solid model, or in principle to liquids, although in the latter case we do not usually have explicit forms for the chemical potential. For an Einstein solid, the logarithm of the partition function is (cf. equation (6.7))

$$\ln Q_s = -3N \ln \left[2 \sinh \left(\frac{\hbar \omega}{2kT} \right) \right] \tag{12.26}$$

Consequently, from equation (12.8), but importantly including a binding energy term,

$$\mu_s = 3kT \ln \left[2 \sinh \left(\frac{\hbar \omega}{2kT} \right) \right] - \frac{1}{2} u_0 z \tag{12.27}$$

where μ_s denotes the chemical potential of the solid referred to a zero energy defined as the energy of an atom at rest *in vacuo*. This accounts for the second term whose form arises in the following way: the binding energy of an isolated pair is u_0, z is the number of nearest neighbours and the factor $\frac{1}{2}$ avoids attributing all the binding energy to one member of the pair. Strictly, this term should have been included in Section 6.2 where it was omitted for simplicity (it does not in any case influence the heat capacity). However, in any example involving mutual equilibrium, it is crucial to ensure that the energy in each phase is referred to the same zero.

For the gas phase, assuming it is to be classical and to obey $pV = NkT$, equation (10.26) can be written in the form

$$\mu_g = -kT \ln \left[\left(\frac{2\pi mkT}{h^2} \right)^{3/2} \frac{kT}{p} \right] \tag{12.28}$$

The spin degeneracy g has been set equal to unity as in equation (12.27). Equating μ_s and μ_g (following equation (12.25)) and rearranging yields

$$p = kT \left(2\pi \frac{mkT}{h^2} \right)^{3/2} \left[\sinh \left(\frac{\hbar \omega}{2kT} \right) \right]^3 \exp \left(-\frac{zu_0}{2kT} \right) \tag{12.29}$$

If g had been included it would have cancelled out in this equation. The Einstein model is not fully satisfactory, as we have seen, and the vapour pressures of solids tend anyway to be very small and difficult to measure. The example was chosen because explicit expressions for chemical potential are available from discussions in earlier chapters. However, a similar approach may be taken to the calculation of vapour pressure of any solid or liquid provided that there is available an expression or measure of the chemical potential in the condensed phase. In the case of a liquid one generally does not have an explicit expression for μ, although it can be related (as in the treatment of the Einstein model above) to the measured vapour pressure by equating μ_g and μ_l. The resulting determination of μ_l is valuable in a discussion of mixtures (see Section 12.3 below). Another useful conclusion is that as $T \to 0$ the vapour pressure becomes vanishingly small and consequently $\mu_g \to 0$. In these circumstances the chemical potential is simply the increase in energy of the system when an atom, initially at rest *in vacuo*, is placed into the liquid under conditions of constant temperature ($T \to 0$) and pressure ($p \to 0$). Thus, as $T \to 0$,

$$\mu_l = -l \qquad (12.30)$$

where l is the latent heat per atom of liquid.

A final point relates to the matter of mixtures of reacting or dissociating gases. These are usually dealt with by the methods described in Section 9.5 and also referred to in Section 4.3 in which the free energy is minimized with respect to the degree of dissociation. However, it is clear that, in principle at least, we could argue that equilibrium is defined by equating the chemical potentials of each type of particle in both possible states (bound or unbound). For example, dissociation in HD proceeds until the chemical potentials of H and D separately are equal in the gaseous form and in the molecular combination. We do not propose to elaborate this, but only note it as yet another example of alternative approaches being possible.

12.3 The ideal mixture

We intend in this section to deal with condensed mixtures. These are in general fairly complicated, but it is the practice to describe measured thermodynamic properties of such mixtures in terms of deviations from an ideal model. The definition of the model is ultimately arbitrary and is found to depend upon one's choice of authority; our definition will be given below. It is usual to proceed from the properties of an ideal gas mixture, as discussed in Section 9.4, where the gases A and B are taken to be ideal (that is, the details of atomic interactions are not important and collisions occur infrequently – see Question 9.1). Letting there be $N_A + N_B$ atoms in volume V, and using the notation

$$X_A = \frac{N_A}{N_A + N_B}; \quad X_B = \frac{N_B}{N_A + N_B} \qquad (12.31)$$

then for component A, assumed classical, equation (12.28) gives

$$\mu_A = -kT \ln \left[\left(\frac{2\pi m_A kT}{h^2} \right)^{3/2} \frac{g_A V}{X_A (N_A + N_B)} \right] \qquad (12.32)$$

$$= -kT \ln \left[\left(\frac{2\pi m_A kT}{h^2} \right)^{3/2} \frac{g_A V}{N_A + N_B} \right] + kT \ln X_A \qquad (12.33)$$

In this expression, only the second term relates to concentration, while the first takes exactly the form which would apply if *all* atoms were of type A. We thus write

$$\mu_A = \mu_A^0 + kT \ln X_A \tag{12.34}$$

Similarly

$$\mu_B = \mu_B^0 + kT \ln X_B \tag{12.35}$$

The Gibbs free energy of this mixture is the simple addition

$$G = N_A \mu_A + N_B \mu_B \tag{12.36}$$

$$= N_A \mu_A^0 + N_B \mu_B^0 + kT[N_A \ln X_A + N_B \ln X_B] \tag{12.37}$$

The final term is, of course, associable with the configurational entropy of mixing arising from counting the number of ways of arranging N_A and N_B among $(N_A + N_B)$ positions. The microcanonical distribution is appropriate here since each possible rearrangement has the same energy in a situation where A-A, A-B and B-B interactions are supposed to be of negligible importance. Thus, using equation (4.14),

$$S = k \ln \Omega \tag{12.38}$$

where

$$\Omega = \frac{(N_A + N_B)!}{N_A! N_B!} \tag{12.39}$$

Hence using Stirling's theorem and equations (12.31), we can write

$$\ln \Omega = -[N_B \ln X_B + N_A \ln X_A] \tag{12.40}$$

Thus, the final term of (12.37) has the form $-TS$ as expected. So much for the classical gas mixture.

Now in making our definition of the ideal *liquid* mixture we argue that the effects of interactions (which are surely important in liquids) would not be expected to affect the entropy of the mixing term significantly. We therefore take as our *definition* the statement that the chemical potentials of each component (labelled α) are given by

$$\mu_\alpha(p, T, X_\alpha) = \mu_\alpha^0(p, T) + kT \ln X_\alpha \tag{12.41}$$

where $\mu_\alpha^0(p, T)$ is the chemical potential of pure component α at the same pressure and temperature as the mixture we refer to. The interaction effects are all lumped into $\mu_\alpha^0(p, T)$. This defined model accurately describes the isotopic gas mixture but is only approximate for real liquid mixtures, largely because of its over-simplification of interactions. Nevertheless, we now use the model with further approximations to derive Raoult's law relating to the composition of the vapour mixtures. This law can be obtained by equating the chemical potentials of each component in vapour and liquid phases. For two components A and B, this means that at a particular temperature and pressure,

$$\mu_A(\text{liq}) = \mu_A(\text{vap}) \tag{12.42}$$

and

$$\mu_B(\text{liq}) = \mu_B(\text{vap}) \tag{12.43}$$

The chemical potentials in the vapour phases can be written explicitly if we assume both components to behave as perfect gases, in which the details of atomic interactions are ignored and Dalton's law of partial pressures (equation (9.33)) is obeyed. We shall denote these partial pressures by p_A and p_B and their limiting values for pure A and B liquids by vapour pressures p_A^0 and p_B^0 respectively. p_A and p_B are functions of concentration and temperature, p_A^0 and p_B^0 only of temperature. Hence, using equation (10.26)

$$\mu_A(\text{vap}) = -kT \ln\left[\left(\frac{2\pi m_A kT}{h^2}\right)^{3/2} \frac{g_A kT}{p_A}\right] \tag{12.44}$$

and

$$\mu_B(\text{vap}) = -kT \ln\left[\left(\frac{2\pi m_B kT}{h^2}\right)^{3/2} \frac{g_B kT}{p_B}\right] \tag{12.45}$$

The forms appropriate for $\mu_A(\text{liq})$ and $\mu_B(\text{liq})$ requires a little more argument. We begin with the definition (12.41):

$$\mu_A(\text{liq}, p, X_A) = \mu_A^0(\text{liq}, p) + kT \ln X_A \tag{12.46}$$

The difficulty is that we do not know what value to take for $\mu_A^0(\text{liq}, p)$. However, if we take p_A^0 (the vapour pressure of pure liquid A) to be known, then $\mu_A^0(\text{liq}, p_A^0)$ can be found by equating it to the chemical potential in the pure A vapour phase:

$$\mu_A^0(\text{liq}, p_A^0) = -kT \ln\left[\left(\frac{2\pi m_A kT}{h^2}\right)^{3/2} \frac{g_A kT}{p_A^0}\right] \tag{12.47}$$

The remaining point is that the difference between $\mu^0(\text{liq}, p)$ and $\mu_A^0(\text{liq}, p_A^0)$ is typically small for moderate pressures and an approximation can be obtained by making a Taylor expansion and using only the first two terms:

$$\mu_A^0(\text{liq}, p) = \mu_A^0(\text{liq}, p_A^0) + (p - p_A^0)\left(\frac{\partial \mu_A^0}{\partial p}\right)_T \tag{12.48}$$

Using equation (12.13) for constant T and taking the liquid to be rather incompressible so that the variation in volume v_A^0 per atom may be neglected, we obtain

$$\mu_A^0(\text{liq}, p) = \mu_A^0(\text{liq}, p_A^0) + (p - p_A^0)v_A^0 \tag{12.49}$$

Thus combining equations (12.46), (12.47) and (12.49),

$$\mu_A(\text{liq}, p, X_A) = -kT \ln\left[\left(\frac{2\pi m_A kT}{h^2}\right)^{3/2} \frac{g_A kT}{p_A^0}\right]$$
$$+ (p - p_A^0)v_A^0 + kT \ln X_A \tag{12.50}$$

Similarly,

$$\mu_B(\text{liq}, p, X_B) = -kT \ln\left[\left(\frac{2\pi m_B kT}{h^2}\right)^{3/2} \frac{g_B kT}{p_B^0}\right]$$
$$+ (p - p_B^0)v_B^0 + kT \ln X_B \tag{12.51}$$

Now, according to equation (12.42) we should equate (12.44) to (12.50), and we may then solve to obtain

$$p_A = X_A p_A^0 \exp\left[\frac{(p - p_A^0)v_A^0}{kT}\right] \tag{12.52}$$

Similarly, equating (12.45) and (12.51),

$$p_B = X_B p_B^0 \exp\left[\frac{(p - p_B^0)v_B^0}{kT}\right] \tag{12.53}$$

Finally

$$p = p_A + p_B = X_A p_A^0 \exp\left[\frac{(p - p_A^0)v_A^0}{kT}\right]$$
$$+ X_B p_B^0 \exp\left[\frac{(p - p_B^0)v_B^0}{kT}\right] \tag{12.54}$$

This is almost in the form of Raoult's law but only becomes exactly so when the exponential factors can be taken as unity. Question 12.2 checks whether that approximation would be valid for a mixture of argon and krypton. We must also remember that another assumption has gone into the derivation of equation (12.54), namely that the vapour phase behaves as an ideal gas mixture. Raoult's law in the form

$$p = X_A p_A^0 + X_B p_B^0 \tag{12.55}$$

is sometimes taken as an alternative definition of the ideal mixture and is not generally consistent with our definition embodied in equation (12.41).

The ideal mixture model, which can readily be generalized to describe any number of components, is commonly used as a comparison for the behaviour of real mixtures or of more sophisticated theoretical models, rather in the way that the Debye model of solids is used as a basis for comparison of real solids. The reader may wish to consult Rowlinson and Swinton[1] for further details and recent references to work on liquid mixtures involving not only vapour pressure but also a range of other thermodynamic parameters.

12.4 Phase separation of mixtures

Many mixtures, both solids and liquids, separate out into distinct phases under certain conditions of temperature and pressure. Examples of this are found in the alloys of Ag/Cu, Pb/Sn, Pb/Sb, Al/Si, Cr/Ni, etc., and among liquids, phenol/water, nicotine/water and of course liquid ^{3}He/^{4}He as discussed in the last chapter. Figure 12.2 shows an example of a phase diagram. The fact of phase separation is broadly accounted for by asserting that the separation will occur if the system can thereby achieve a lower free energy, otherwise not. This follows from our argument that equilibrium is characterized by a minimized free energy. As a highly simplified illustration of this approach let us consider the stability of a solid mixture compared with that of the separate pure components A and B. If we are to understand this, it is most important that the inter-atomic binding forces be explicitly considered, because the ideal mixture of Section 12.3 would not be expected ever to phase-separate. The whole matter hinges on whether atoms A prefer (energetically) the company of A or B atoms, and to what quantitative extent. We shall simplify matters by concentrating on the energy aspects, reducing the significance of entropy terms TS by considering the low-temperature limit $T \to 0$. This is the opposite of our approach in the last section where the entropy terms were of considerable importance. There is further advantage of taking the limit $T \to 0$ which is that thermal vibrations in the solid do not then contribute to the chemical potential and, relative to an atom isolated *in vacuo*, μ is given simply by minus the binding energy of

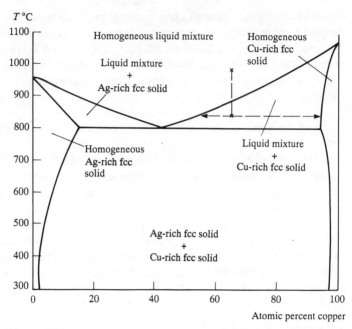

Figure 12.2 Phase-separation diagram for the alloy system copper/silver. As an example of how to 'read' this diagram, suppose we take a liquid mixture with 65 % copper near 1000 °C and cool it to 850 °C as indicated by the vertical dashed line. The resulting material would be, according to the diagram, heterogeneous, with solid 95 % copper in equilibrium with liquid 55 % copper. The reader may wish to try some other possible processes. The procedure is that, if the resulting material is heterogeneous, the two concentrations are obtained by drawing a horizontal isotherm to cross the boundaries of the heterogeneous area of the phase-separation diagram.

the atom in the condensed state. This statement was also embodied in equation (12.27). In the mixture, let X_A denote the relative concentration of component A and let z be the number of nearest neighbours appropriate to the structure of the mixture (we ignore all but nearest neighbours here). Then on average (assuming random distribution) there are zX_A atoms of type A next to any given atom and zX_B of type B ($X_A + X_B \equiv 1$). If there are N atoms in all, then NX_A are of type A and NX_B are of type B. Thus the numbers N_{AA}, N_{BB} and N_{AB} of pairs of neighbours of types AA, BB and AB respectively are given by

$$N_{AA} = \frac{1}{2}NzX_A^2 \qquad N_{BB} = \frac{1}{2}NzX_B^2 \qquad N_{AB} = NzX_AX_B \quad (12.56)$$

where the factors $\frac{1}{2}$ avoid double counting.

If the bond energies are U_{AA}, U_{BB} and U_{AB}, then (ignoring entropy and thermal vibration terms as mentioned above)

$$G = N_{AA}U_{AA} + N_{BB}U_{BB} + N_{AB}U_{AB} \tag{12.57}$$

$$= \tfrac{1}{2} Nz [X_A U_{AA} + X_B U_{BB} + X_A X_B (2U_{AB} - U_{AA} - U_{BB})] \tag{12.58}$$

The first two terms give the free energies of the crystals of pure components A and B and would be the only contributions in the completely separated pure phases (characterised by $(X_A = 1, X_B = 0)$ or $(X_A = 0, X_B = 1)$). So the whole matter of whether separation into two phases, not in general pure but with each characterized by different (X_A, X_B), is energetically preferred rests on the sign of the bracket $(2U_{AB} - U_{AA} - U_{BB})$; if negative the mixture is energetically favoured, if positive it is not. Since the bracket can in practice take either sign, we have established the possibility of phase separation.

We will now concentrate on how to calculate the equilibrium concentrations in the two phases assuming phase separation to have occurred. The approach will be as follows: consider an alloy with two atomic varieties A and B, which has phase-separated into phases 1 and 2. A complete solution of the problem will be given by results for the concentration $X_A(1)$ and $X_A(2)$. The *quantities* of each phase are simply governed by conservation of the total number of atoms of each variety. The concentrations $X_A(1)$ and $X_A(2)$ can in principle be obtained from the two equations which equate the chemical potentials of each variety in each phase

$$\mu_A(1) = \mu_A(2) \tag{12.59}$$

$$\mu_B(1) = \mu_B(2) \tag{12.60}$$

This method can be applied to crystalline alloys or to liquid mixtures. A rather nice example of this is the case of phase separation in liquid ^{3}He/^{4}He mixtures mentioned in the last chapter, and the reader should refer to Section 11.5. Figure 12.3 shows the phase diagram, and we shall now attempt to account for the details as $T \to 0$, particularly the origin of the remaining 6 % of pure ^{3}He. Figure 12.4 shows two intersecting curves. The first, the continuous line, is a plot from equation (11.42) of the Fermi degeneracy temperature $T_f(X_3)$ of the dissolved ^{3}He in a dilute solution of ^{3}He in liquid ^{4}He. It is assumed here, as discussed in Section 11.5, that the ^{3}He atoms behave as a perfect Fermi gas.

The second (dotted) line arises as follows. The chemical potential of a perfect Fermi gas approaches kT_f as $T \to 0$ but since we are here dealing with ^{3}He atoms in liquid solution, rather than strictly in the gas phase, we must include an extra term $\epsilon_3(X_3)$ representing the binding energy of a ^{3}He atom in a predominantly liquid-^{4}He environment. Thus, using $\mu_3(4)$ to denote the chemical potential of the ^{3}He atoms in dilute solution, we write

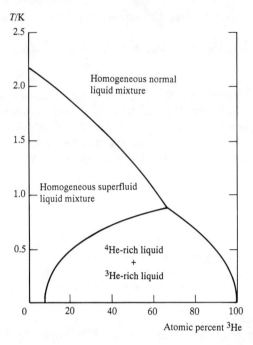

Figure 12.3 Phase-separation diagram for the liquid system of helium isotope mixtures (^{3}He + ^{4}He). The interpretation of the diagram is exactly as for Figure 12.2. The diagram is from Wilks and Betts (1987) *An Introduction to Liquid Helium*, 2nd ed. (Oxford, 1987), where detailed references can be found. Above 0.87 K liquid ^{3}He and ^{4}He are miscible in all proportions, but at lower temperatures a phase separation occurs. The concentration is denoted by $X_3 = N_3/(N_3 + N_4)$ where N_3 and N_4 are the numbers of ^{3}He and ^{4}He atoms respectively. The λ-point of pure ^{4}He ($X_3 = 0$) is at 2.17 K but the addition of ^{3}He lowers T_λ until the tricritical point is reached at 0.87 K with $X_3 = 0.67$. The lambda line and the phase separation lines divide the phase diagram into three regions: (i) essentially normal liquid ^{3}He/^{4}He mixture; (ii) essentially superfluid liquid ^{4}He with some dissolved ^{3}He; and (iii) a two-phase region in which the liquid separates into ^{3}He-rich and ^{4}He-rich solutions. As the temperature is reduced towards the absolute zero, the lighter phase consists of almost pure ^{3}He, while the heavier ^{4}He-rich phase contains about 6 % of ^{3}He.

$$\mu_3(4) = kT_f(X_3) - \epsilon_3(X_3) \tag{12.61}$$

Accepting Figure 12.3 we have to understand why the dilute solution of ^{3}He in liquid ^{4}He, with X_3 equal to a few percent can be in equilibrium with a phase with a ^{3}He concentration close to 100 %. We need an expression for the chemical potential of ^{3}He atoms in an almost pure ^{3}He environment, $\mu_3(3)$, and this is just $-l_3$, where l_3 is the latent heat per atom of liquid ^{3}He. This follows from equation (12.30) since we are in fact considering the case $T \to 0$. Thus

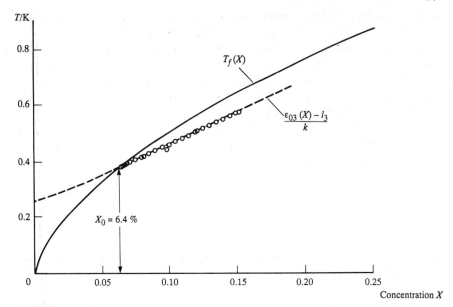

Figure 12.4 The method used to obtain $X_0 = 6.4\%$ from equation (12.63). The experimental circles are due to (a) D.O. Edwards, D.F. Brewer, P. Seligman, M. Skertic, and M. Yaqub, *Phys. Rev. Lett.* (1965), **15**, 773, (b) E.M. Ifft, D.O. Edwards, R.E. Sarwinski, and M.M. Skertic, *Phys. Rev. Lett.* (1967), **19**, 831, and (c) D.O. Edwards, E.M. Ifft, and R.E. Sarwinski, *Phys. Rev.* (1969), **177**, 380; and $T_f(X)$ is calculated from equation (11.42). The point of intersection yields X_0 at $T = 0$.

$$\mu_3(3) = -l_3 \tag{12.62}$$

Equating chemical potentials $\mu_3(4) = \mu_3(3)$ now gives, after division by k,

$$T_f(X_3) = \frac{\epsilon_3(X_3) - l_3}{k} \tag{12.63}$$

The second line in Figure 12.4 shows experimental data for the right-hand side of equation (12.63), independently obtainable from subsidiary experiments. $\epsilon_3(X_3)$ could be obtained from ^{3}He partial vapour pressure measurements on an X_3 liquid mixture at finite temperatures, equating chemical potentials as in the last section, and extrapolating the results towards $T = 0$ (actually, it is not quite as straightforward as it sounds, but the principle is right). l_3 is equal to the latent heat per atom of pure liquid ^{3}He, extrapolated towards $T = 0$. Figure 12.4 illustrates the procedure. Problems arise in this procedure because $\epsilon_3(X_3)$ and l_3 are not very different in magnitude, and equation (12.63) requires their difference to be known with as much precision as possible. However, the details[2,3] need not concern us here and we simply take it that the left-hand side of equation (12.63) is calculable from equation (11.2) and the right-hand side is obtainable from independent experiments. Thus, equation (12.63) may be solved for X_3 as $T \to 0$, and the possible existence of a limiting value was predicted[2] before the details of the phase diagram had been accurately established.[4] The limiting value of $X_3 = 6.4\%$ is known to be consistent with both equation (12.63) and experiment.

The right-hand corner of the phase diagram may be accounted for in a similar way. The difference is that ^{4}He atoms in ^{3}He solution are never dense enough as $T \to 0$ to show Bose degeneracy, so that we use the classical form (10.28) for the chemical potential:

$$\mu_4(3) = -\epsilon_4(3) - kT \ln \left[\left(\frac{2\pi m_4 kT}{h^2} \right)^{3/2} \frac{2V}{N_4} \right] \quad (12.64)$$

where $\epsilon_4(3)$ is the binding energy of a ^{4}He atom in a predominantly ^{3}He environment. Also, from equation (12.30),

$$\mu_4(4) = -l_4 \quad (12.65)$$

where l_4 is the latent heat per atom of liquid ^{4}He. Equating chemical potentials $\mu_4(3) = \mu_4(4)$ now gives

$$l_4 = \epsilon_4(3) + kT \ln \left[\left(\frac{2\pi m_4 kT}{h^2} \right)^{3/2} \frac{2V}{N_4} \right] \quad (12.66)$$

Substituting numerical values and remembering that N_4/V is proportional to X_4, equation (12.66) predicts that

$$X_4 = 0.85 T^{3/2} \exp\left(-\frac{0.56}{T}\right) \quad (12.67)$$

The remarkable success of this prediction when compared with direct experimental measurements is illustrated in Figure 12.5.

12.5 Osmotic pressure

When two solutions of different concentrations are separated by a diathermic semi-permeable barrier which will transmit A atoms freely but holds back B atoms, a pressure difference is observed. This is known as osmotic pressure, usually denoted by Π. Figure 12.6 shows a possible experimental apparatus, and we shall consider the situation in which one side of the barrier contains pure liquid A, and the other side contains a mixture with concentrations X_A and X_B.

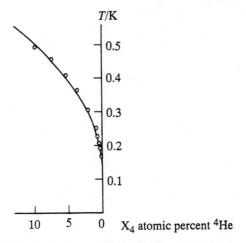

Figure 12.5 The right-hand corner of the phase diagram shown in Figure 12.3. The data is due to J.P. Laheurte and J.R.G. Keyston, *Cryogenics* 11 (1971), 485, and the continuous line has the form of equation (12.67). The agreement is seen to be good.

We shall use the ideal mixture model as described in Section 12.3, which is most likely to be appropriate when X_B is small. The barrier blocks the passage of B atoms so that we cannot equate *their* chemical potentials. However, A atoms, which do reach equilibrium concentrations,

$$\mu_A^0(p, T) = \mu_A(p + \Pi, T, X_A) \tag{12.68}$$

where Π is the osmotic pressure and X_A ($=1 - X_B$) is the concentration of A on the right-hand side of Figure 12.6. Assuming the definition of an ideal mixture, equation (12.41), we now write

$$\mu_A^0(p + \Pi, T, X_A) = \mu_A^0(p + \Pi, T) + kT \ln X_A \tag{12.69}$$

Now we expand $\mu_A^0(p + \Pi, T)$ about the pressure p, taking only the first two terms:

$$\mu_A^0(p + \Pi, T) = \mu_A^0(p, T) + \Pi \left(\frac{\partial \mu_A^0}{\partial p} \right)_T \tag{12.70}$$

In using ideal mixture equations, we should strictly limit ourselves to concentrations X_A close to 100 %. Equation (12.13) gives

$$\left(\frac{\partial \mu_A^0}{\partial p} \right)_T = v_A^0 \tag{12.71}$$

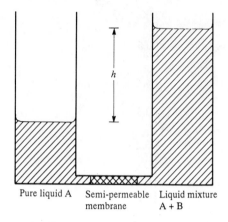

Pure liquid A   Semi-permeable   Liquid mixture
                 membrane   A + B

Figure 12.6 Notional arrangement for measuring the osmotic pressure due to the solute B relative to pure liquid A. The semi-permeable membrane is permeable to A but not to B.

and for most liquids v_A^0 varies little with pressure or with small X_B. Hence

$$\mu_A^0(p + \Pi, T) = \mu_A^0(p, T) + \Pi v_A^0 \tag{12.72}$$

and thus from equation (12.69)

$$\mu_A(p + \Pi, T, X_A) = \mu_A^0(p, T) + \Pi v_A^0 + kT \ln X_A \tag{12.73}$$

Finally, from equation (12.68) equating chemical potentials, we obtain

$$\Pi v_A^0 = -kT \ln X_A \tag{12.74}$$

$$= -kT \ln(1 - X_B) \tag{12.75}$$

Thus, again taking X_B as small, and expanding the logarithm,

$$\Pi v_A^0 = X_B kT \tag{12.76}$$

This relation is called 'van't Hoff's formula' and it is applicable to weak solutions independently of the particular solvent and solute concerned. It is strongly reminiscent of the equation of state of an ideal gas, with the gas pressure replaced by osmotic pressure, gas volume by solution volume, and the number of particles in the gas by the number of atoms (or molecules) of solute. It is a good description of the behaviour of real solutions including ^{3}He/^{4}He mixtures discussed in the last section, provided that $T \gg T_f$. In fact, as in other examples, these mixtures with ^{3}He as solute (B) and superfluid liquid ^{4}He as solvent (A) provide a particularly pleasing example of this effect. The superfluidity of liquid ^{4}He makes it easy to construct a semi-permeable membrane ('a superleak') which passes ^{4}He but blocks ^{3}He. When $T \gg T_f$, van't Hoff's formula is found[5] to be correct for a range of concentrations X_3. When $T \ll T_f$ equation (12.76) has to be replaced by a suitable Fermi form which we do not propose to discuss here.

Osmotic pressure can be quite large and should not be regarded as an esoteric matter. It plays a vital role in the circulation of fluids in trees and other plants.

12.6 Electron paramagnetism

We treat the valency electrons as a degenerate Fermi gas as in Section 11.2. Now electrons have a spin and, since they are charged, a magnetic moment $e\hbar/2m$ which, of two allowed, we shall denote by γ. In a magnetic field B each electron may align itself in one or two directions, parallel (↑) or antiparallel (↓) with the direction of B. As $B \to 0$ the numbers opting for each of the possibilities approach equality, and in general we may regard the electron gas as a mixture of two different species distinguishable by the orientation (↑ or ↓). When $B = 0$ there is no energy difference between the two possible orientations and it would be natural to use equation (10.32) for the Fermi energy (chemical potential) ϵ_f,

$$\epsilon_f = \frac{h^2}{8m} \left(\frac{3N}{\pi V} \right)^{2/3} \tag{12.77}$$

The density of states for each 'subgas' is then a half of the expression for the whole gas, that is

$$g(\epsilon) = \frac{2\pi V}{h^3} (2m)^{3/2} \epsilon^{1/2} \tag{12.78}$$

This function is represented diagrammatically in Figure 12.7(a).

Also shown in Figure 12.7(a) are the Fermi energies $\epsilon_{f\uparrow}$ and $\epsilon_{f\downarrow}$ indicating how far up the ladder of levels electrons reach at one per level. Figure 12.7(b) shows what happens instantaneously when a field B is applied. The potential energies of the subgases shift by $\pm\gamma B$ relative to the original zero of energy so that one side of the diagram moves down by γB, the other side up by the same amount. Clearly this is not an equilibrium situation because the Fermi energies (chemical potentials), referred to the original zero, have become unequal:

$$\epsilon_{f\uparrow} = \epsilon_f + \gamma B; \qquad \epsilon_{f\downarrow} = \epsilon_f - \gamma B \tag{12.79}$$

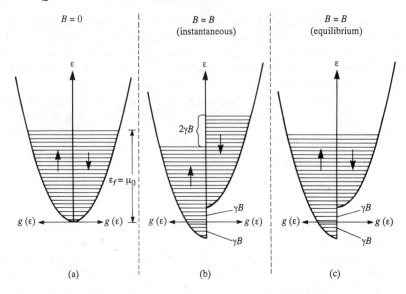

Figure 12.7 (a) The parabola represents $g(\epsilon)$ given by equation (12.78), with distinction made between electrons with spin-up (↑) and those with spin-down (↓). Both varieties occupy states up to the Fermi energy $\epsilon_f \equiv \mu_0$. (b) Since spinning electrons have associated magnetic moments, the switching on of a magnetic field instantaneously shifts the whole ladder of energy levels by $\pm \gamma B$, that is, oppositely for the two varieties. (c) After equilibrium is established by some electrons tipping from right to left by flipping their spins, there are more with spin-up so that there is a net magnetic moment. The diagrams are exaggerated in the sense that for accessible fields it is usually the case that $\gamma B \ll \epsilon_f$.

Consequently, as equilibrium is approached, electrons reverse their orientation, spilling from right to left in Figure 12.7(b) until the Fermi level (chemical potential) equalizes on the two sides, Figure 12.7(c). The question is, how many electrons ΔN have to reverse their orientation to achieve this? For moderate field B, it is a relatively small number which can be calculated as follows. The number of states (electrons) between the unequal levels in Figure 12.7(b) is the product of the level difference $2\gamma B$ and the density of states $g(\epsilon_f)$ in that region, i.e., $2\gamma B g(\epsilon_f)$. The number of electrons which have to change sides to bring the levels together is thus one half of this, i.e.,

$$\Delta N = \gamma B g(\epsilon_f) = \gamma B \frac{2\pi V}{h^3} (2m)^{3/2} \epsilon_f^{1/2} \tag{12.80}$$

But

$$\epsilon_f^{3/2} = \frac{h^3}{8(2m)^{3/2}} \left(\frac{3N}{\pi N} \right) \tag{12.81}$$

Hence

$$\Delta N = \gamma B \frac{3N}{4\epsilon_f} \qquad (12.82)$$

The magnetic moment is thus

$$(\tfrac{1}{2}N + \Delta N)\gamma + (\tfrac{1}{2}N - \Delta N)(-\gamma) = 2\gamma \Delta N \qquad (12.83)$$

$$= \gamma^2 B \frac{3N}{2\epsilon_f} \qquad (12.84)$$

Thus the magnetic susceptibility $\chi = \mu_0 M/VB$ is given by

$$\chi = \frac{3}{2}\left(\frac{\mu_0 \gamma^2}{\epsilon_f}\right)\frac{N}{V} = \text{constant} \qquad (12.85)$$

where μ_0 is here the permeability (in SI units) of free space.

In fact, although we have only considered one source of magnetism in metals, equation (12.85) is certainly an appropriate contribution which can be experimentally distinguished. However, the situation is complicated and for further details the reader is referred elsewhere.[6]

An exactly similar approach can be taken to pure liquid ^{3}He or to ^{3}He/^{4}He solutions since the ^{3}He has a nuclear magnetic moment (^{4}He does not). The nuclear magnetic susceptibility of dilute solutions of ^{3}He in liquid ^{4}He is particularly interesting because, as explained elsewhere, it is possible experimentally to follow the transition from classical to Fermi degenerate behaviour. At low temperatures ($T \ll T_f$) equation (12.85) is obeyed, and as the temperature is raised there is a transition to the classical Curie law behaviour when $T \gg T_f$,

$$\chi = \frac{\mu_0 \gamma^2}{kT} \frac{N}{V} \qquad (12.86)$$

The reader is invited in Question 12.9 to prove equation (12.86). Experiments agree well with predictions.[7]

It might be felt that ^{3}He/^{4}He solutions have been used too often as illustrations – of phase-separation, osmotic pressure, Fermi degeneracy, heat capacity and magnetic susceptibility. We have been quite deliberate about this, since this particular system shows a range of effects in an unusually simple and uncluttered way and has been thoroughly investigated.

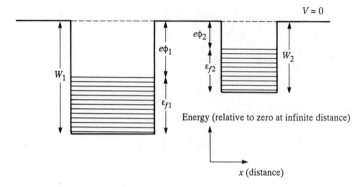

Figure 12.8 Energy diagram for two metals *before* contact is established. Actually this represents a non-equilibrium situation since electrons will tend to move from metal 2 to metal 1 through the intervening electron vapour, but this process would take an extremely long time.

12.7 Contact potential

When two dissimilar metals are in equilibrium with each other, there is a potential difference between them. The equilibrium could possibly be reached through the electron gas between the metals but, since this gas density is minute, the attainment of equilibrium would take a very long time. It is easier in practice to establish equilibrium by placing the metals in conducting contact either directly or through a wire. It is then found that a potential difference exists between the metals. This potential difference can be measured by suitable techniques[8] and its explanation has to do with the equalization of Fermi energies (chemical potentials) in the two metals. Figure 12.8 shows diagrammatically the various energies for each metal *before* equilibrium contact is achieved.

Firstly, there is the depth W of the potential well, that is, the amount by which the energy of a single-particle ground-state free electron is below that of an electron at rest at an infinite distance away. Secondly, there is the Fermi energy ϵ_f, that is, the energy of an electron at the top of the Fermi distribution relative to a ground-state electron. Thirdly, there is the work function Φ (usually measured in volts) where $e\Phi$ is the energy to remove an electron from the top of the Fermi distribution and take it to infinity. These energies are related by

$$W = \epsilon_f + e\Phi \qquad (12.87)$$

In general, the Fermi levels are not the same in both metals, relative to the common energy zero at infinity. Thus, when contact is made, electrons will flow so as to equalize the levels (chemical potentials). The situation

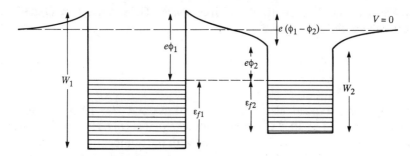

Figure 12.9 Energy diagram for two metals *after* electrical contact is established. Charge passes as discussed in the text and as a result, external fields are set up and a contact potential exists.

seems rather like that of electron paramagnetism discussed in the last section. The difference lies in the crucial point that when charged electrons leave or enter a metal they change its potential so that W changes as well as ϵ_f. Figure 12.9 shows the result, with all changes producing a balance of Fermi levels. It is easy to show that the actual flow of electrons is so slight that ϵ_f would be barely affected, even if the surfeit or depletion of electrons were spread evenly over the bulk of the metals. In fact the ordinary arguments of electrostatics show that it is energetically favourable for the electronic number density to be completely unaltered in the bulk with any charge imbalance being accommodated in a very thin surface layer. Consequently, the dominant effect is the relative shift in W of the two metals, which may be visualized, as in Figure 12.9, as a relative shift of the two boxes. The relative energy shift required to bring the Fermi levels together is just $e\Delta V$ where ΔV is the contact potential.

Thus

$$e\Delta V = (W_1 - \epsilon_{f1}) - (W_2 - \epsilon_{f2}) \tag{12.88}$$

$$= e\Phi_1 - e\Phi_2 \tag{12.89}$$

from equation (12.87). Hence

$$\Delta V = \Phi_1 - \Phi_2 \tag{12.90}$$

Experimentally, contact potentials are typically in the range 0–4 volts. Let us take a particular pair of metals (Pt/Cu, for which $\Delta V = 0.91$ volts) and specify their shapes as flat plates of area A facing each other and separated by d as in a parallel plate condenser. The capacitance of the arrangement is

$$C = \frac{\epsilon_0 A}{d} \tag{12.91}$$

where ϵ_0 is the SI permittivity of free space. Thus, since $C \equiv Q/\Delta V$, the transferred charge is

$$Q = \frac{\epsilon_0 A \Delta V}{d} \qquad (12.92)$$

Taking some typical sizes, let $A = 0.01 \text{ m}^2$ and $d = 10 \text{ mm}$. Then the charge Q is 1.14×10^{-6} Coulombs which equates to 7.15×10^{12} electrons. This number is of course tiny compared with the number of electrons in, say, 1 kg of copper (9.4×10^{24}) and is in any case concentrated on the surface, as mentioned above, so that the Fermi energies (relative to their respective single-particle ground-state energies) remain unaltered in the two metals. This example is included because the charged nature of the electrons is a major factor.

12.8 Surfaces

The study of surfaces has become increasingly important in recent years because of the fact that surface effects often have applications with industrial and economic consequences. In the chemical industry, for example, catalysis at surfaces significantly accelerates some reactions. And in the electronics industry, the importance of semiconductor devices and their miniaturization is enormous; often specialized devices are manufactured atomic layer by atomic layer and the techniques for doing this obviously depend on a good knowledge of the physics and chemistry involved. Our intention here is to present a few important ideas relating to the statistical mechanics of these types of systems.

The picture normally includes a *substrate*, that is, a volume of solid or liquid with its own surface. In our brief discussion we will be concerned primarily with approximately smooth solid substrates and we shall further limit the discussion by focusing attention on the properties of adsorbed layers on the substrate surface rather than on the substrate surface itself. The substrate may be crystalline (but with a structure which may differ from that in the bulk) or amorphous. Adsorption can be a physical effect, explicable in terms of simple van der Waals forces between the atoms or molecules and the substrate, or it may involve chemical bonding. Adsorbed atoms or molecules interact of course with the substrate and we can think in terms of a binding energy and, at least for some purposes, adsorption sites. Adsorbed atoms or molecules also interact with each other, and there is evidence to show that the phase diagrams of adsorbed layers can be quite complicated, with two-dimensional (or restricted geometry) gas, liquid and solid phases. At low temperatures, quantum statistics may become important, and experimental techniques allow the possibility of measuring some thermodynamic properties such as heat capacity and magnetic susceptibility. Further, if the layers are electrically conducting, then a range of special effects can be seen at low temperatures, including superconductivity and the quantum Hall effect. There is thus a vast menu to choose from and we choose just three relatively simple items.

12.9 Adsorption isotherms

Adsorbed layers of atoms or molecules are of course in thermal equilibrium with the substrate, and with the phase above the layer. We will deal here only with the relatively easy case of a gas phase, whose chemical potential takes the form of equation (10.26). The chemical potential of the adsorbed atoms must be equal to this, but the nature of the adsorbed phase depends sensitively in practice on the characteristics of the substrate and the adsorbate, and on the number of adsorbed molecules per unit area of surface. We shall consider an example of *localized* adsorption and another of *mobile* adsorption. In the former, molecules are held in potential wells between atoms at distinguishable sites on the surface of the substrate, while in the latter the molecules are bound to remain close to the surface of the substrate but can move more or less freely in two dimensions in a gas-like or liquid-like manner.

First we discuss a semi-realistic model of multilayer localized adsorption. The substrate is supposed to provide M equivalent, independent, and distinguishable sites on which any number s, from zero to a maximum m, of molecules can be bound in a sort of one-dimensional stack. These sites correspond to potential wells on an array with a spacing of approximately atomic dimensions. Each site has a subsystem canonical partition function $q(s)$ depending, obviously, on the energy states available to the stack of s molecules. If there is a total of N molecules bound on the M sites, and if the number of sites having s molecules is a_s, then all allowed microstates within the model must satisfy two conditions,

$$\sum_{s=0}^{m} a_s = M \quad \text{and} \quad \sum_{s=0}^{m} s a_s = N \qquad (12.93)$$

The canonical partition function for the system of M sites is found by adding a term for each allowed microstate. The form of the term corresponding to one microstate, with each distinguishable site i having its own value of s, is simply (see Appendix 3) the product of all the site partition functions, that is,

$$\prod_i q_i = (q(0))^{a_0} (q(1))^{a_1} \ldots (q(m))^{a_m} \qquad (12.94)$$

We now have to multiply by the number of ways in which the same set of values of a_s may be distributed among the M sites, giving

$$\frac{M!(q(0))^{a_0}(q(1))^{a_1}(q(2))^{a_2}\ldots(q(m))^{a_m}}{a_0!a_1!\ldots a_m!} \tag{12.95}$$

Finally, we sum over all sets of values of a_s consistent with the constraints (12.93), i.e.,

$$Q = \sum_{\{a_s\}}{}'' \frac{M!(q(0))^{a_0}(q(1))^{a_1}\ldots(q(m))^{a_m}}{a_0!a_1!\ldots a_m!} \tag{12.96}$$

where $\{a_s\}$ denotes a set of a_s, and the primes on the summation are reminders of the two constraints of (12.93). Equation (12.96) is not easy to manipulate and it is more convenient (as for Fermi and Bose gases in Section 10.1) to pass to the grand canonical partition function. Since there are M sites each of which can accommodate a maximum of m of molecules, there are mM molecules altogether, so

$$\Xi = \sum_{N=0}^{mM} Q(N, M, T)\lambda^N \tag{12.97}$$

$$= \sum_{\{a_s\}}{}' \frac{M!(q(0)\lambda^0)^{a_0}(q(1)\lambda^1)^{a_1}(q(2)\lambda^2)^{a_2}\ldots(q(m)\lambda^m)^{a_m}}{a_0!a_1!\ldots a_m!} \tag{12.98}$$

where $\lambda \equiv \exp(\mu/kT)$ and the single prime on the summation (12.98) indicates that only the first constraint now needs attention. In the last step we have used the second constraint of (12.93) to factorize λ^N as

$$\lambda^N = \lambda^{\Sigma s a_s} = \lambda^{0 \times a_0} \times \lambda^{1 \times a_1} \times \lambda^{2 \times a_2} \times \ldots \times \lambda^{m \times a_m} \times \ldots$$

Now equation (12.98) can be factorized by the multinomial theorem:

$$\Xi(\mu, M, T) = (\xi(\mu, T))^M \tag{12.99}$$

where

$$\xi = q(0)\lambda^0 + q(1)\lambda^1 + q(2)\lambda^2 \ldots + q(m)\lambda^m \tag{12.100}$$

Now we can use equation (10.10) to obtain the mean site occupation number $\bar{s}$:

$$\bar{s} = kT\left(\frac{\partial \ln\xi}{\partial \mu}\right)_T \tag{12.101}$$

We can now easily derive the Brunauer–Emmett–Teller (BET) equation for multimolecular adsorption. It is merely a special case in which m is taken as infinite, $q(0) = 1$ and $q(s) = q_1 q_2^{s-1}$ where the subscripts label the positions of molecules in the stack. Physically this is saying that the bottom molecule in the stack ('first layer', $s = 1$) has partition function q_1 while all others in the stack have q_2. The model is not perhaps entirely convincing but the prediction is algebraically explicit and proves a useful way of describing real adsorption isotherms. The algebraic simplifications lead to

$$\xi = 1 + q_1 \lambda (1 + (q_2 \lambda) + (q_2 \lambda)^2 + \ldots) \tag{12.102}$$

The infinite sum has an explicit form and the result is

$$\xi = \frac{1 + (q_1 - q_2)\lambda}{1 - q_2 \lambda} = \frac{1 + (q_1 - q_2)\exp(\mu/kT)}{1 - q_2 \exp(\mu/kT)} \tag{12.103}$$

The final step is to use equation (12.101) to obtain

$$\bar{s} = \frac{cx}{(1 - x + cx)(1 - x)} \tag{12.104}$$

where $c = q_1/q_2$ and $x = q_2 \exp(\mu/kT)$, both being functions of temperature. Now the chemical potential must be the same in the adsorbate as in the vapour (gas) and we know from equation (10.25) that $\exp(\mu/kT)$ is inversely proportional to the density V/N and so proportional to the pressure p. We use the physical observation that as p approaches the bulk vapour pressure p_0, the layer thickness, proportional to $\bar{s}$, will become very large; hence we may identify x as p/p_0. Now $\bar{s}$ is proportional, in an experiment, to the amount of adsorbate allowed to condense on the substrate. For these reasons, data tend to be plotted as amounts versus p/p_0. As an illustration, consider a situation in which $c \gg 1$, say, $c = 100$. Figure 12.10(a) shows a plot of $\bar{s}$ versus x from equation (12.104) and one can see that the first layer is almost completely filled before higher layers begin. Also shown, in Figure 12.10(b), are some actual data showing very similar behaviour.

12.10 Two-dimensional solid layers

The adsorbed layer may be solid or fluid, depending on temperature and other parameters. Let us consider the solid case first by discussing the two-dimensional Debye model. The justification for this is that in the direction perpendicular to the substrate surface, the vibration frequency will be relatively large so that modes much above the ground state will not be important – particularly at low temperatures. We refer the reader to Section 6.5

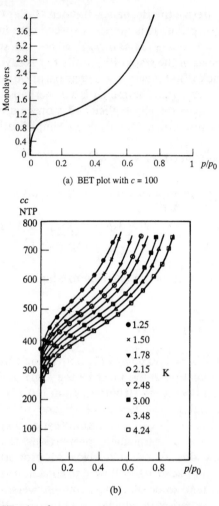

Figure 12.10 (a) BET plot from equation (12.104) with $c = 100$. (b) Adsorption isotherms of ^{4}He on ^{TM}Vycor porous glass at several different temperatures; figure adapted from D.F. Brewer, Chapter 6 entitled 'Multilayer Adsorbed Helium in Restricted Geometries' of *The Physics of Liquid and Solid Helium, Part II*, edited by K.H. Bennemann and J.B. Ketterson (Wiley-Interscience, 1978).

dealing with the normal three-dimensional case, and give the two-dimensional equivalents to equations (6.46) and following, using the same notation. The wave equation for a small displacement u is

$$\frac{\partial^2 u}{\partial x^2} + \frac{\partial^2 u}{\partial y^2} = \frac{1}{c^2}\frac{\partial^2 u}{\partial t^2} \tag{12.105}$$

Considering for convenience a square of area L^2 whose walls are held laterally motionless, the appropriate solution is

$$u = \text{const.} \sin\left(n_x \frac{\pi x}{L}\right) \sin\left(n_y \frac{\pi y}{L}\right) \cos\omega t \qquad (12.106)$$

where n_x and n_y are non-zero positive integers and

$$n_x^2 + n_y^2 = \frac{L^2 \omega^2}{\pi^2 c^2} \qquad (12.107)$$

This equation in fact describes a circle of radius $L\omega/\pi c$ in (n_x, n_y)-space, in which unit area represents one mode. Thus the number of modes with frequencies between zero and ω is simply the quadrant (because n_x, n_y must be positive) area of the circle, i.e.,

$$\frac{\pi L^2 \omega^2}{4\pi^2 c^2} = \frac{A\omega^2}{4\pi c^2} \qquad (12.108)$$

Hence by differentiating and multiplying by two to take account of the fact that the wave-equation might describe a compressional displacement or a transverse displacement, we obtain the frequency spectrum

$$g(\omega) = \frac{A\omega}{\pi c^2} \qquad (12.109)$$

Finally, since there must be only $2N$ modes in all, we require

$$\int_0^{\omega_m} g(\omega) \, d\omega = 2N \qquad (12.110)$$

Equations (12.109) and (12.110) combine to give

$$g(\omega) = 4N \frac{\omega}{\omega_m^2} \qquad (12.111)$$

Now we can evaluate $\ln Q$ by integration as in the three-dimensional case:

$$\ln Q = -\frac{4N}{\omega_m^2} \int_0^{\omega_m} \omega \ln\left[2\sinh\left(\frac{\hbar\omega}{2kT}\right)\right] d\omega \qquad (12.112)$$

which leads in the standard way to

$$\frac{C_V}{Nk} = \frac{4}{x_m^2} \int_0^{x_m} \left(\frac{1}{2}x\right)^2 \text{cosech}^2\left(\frac{1}{2}x\right) x \, dx \qquad (12.113)$$

where $x \equiv \hbar\omega/kT$ and $x_m \equiv \hbar\omega_m/kT$. It is now easy to show that

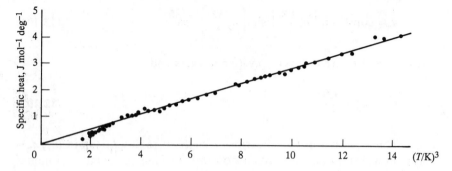

Figure 12.11 Specific heat of a sub-monolayer of ^{3}He on ™Vycor porous glass, plotted against the square of the temperature; figure adapted from D.F. Brewer, A. Evenson, and A.L. Thomson, *J. Low Temp. Phys.* (1970), **3**, 603.

$$C_V = 2Nk \qquad \text{at high } T, \ (x_m \ll 1) \qquad (12.114)$$

and

$$C_V = 24Nk\zeta(3)\left(\frac{kT}{\hbar\omega_m}\right)^2 \quad \text{at low } T, \ (x_m \gg 1) \qquad (12.115)$$

Experiments can be designed which demonstrate this predicted result. One can see two important elements in the design by considering a solid substrate with N_s molecules and an adsorbed layer with N_a molecules of a different type. The heat capacity of the substrate itself will tend to dominate at ordinary laboratory temperatures simply because it is likely to have far more molecules. So the first feature of a successful experiment is to use a porous material for the substrate so that there is plenty of surface area per unit volume. Ideally, the pores should be very fine; for example if one thinks of the simple geometry of a hexagonal close-packed assembly of spheres of diameter d, the surface per unit volume is $3\pi/d\sqrt{2}$ from which it is clear that d should be as small as possible in order to make the contribution of the surface layer to the heat capacity as large as possible. The second element in the design arises from the fact that at low temperatures the heat capacity of the substrate (with N_3 molecules and a constant C_3) takes the Debye form $C_3 N_3 T^3$ whereas that of the adsorbate (with N_2 molecules and a constant C_2) is expected to take the form of equation (12.115), i.e., $C_2 N_2 T^2$. As the temperature falls, for given N_2 and N_3, the ratio of the second to the first of these two terms rises as $1/T$, although of course their sum falls, making the experiment more subject to scatter. So the experiments are done with a porous substrate at temperatures which are low enough to allow the surface contribution to the heat capacity to be significant but not so low that

12.11 Two-dimensional gaseous layers

It turns out that the analyses of perfect gases of bosons and fermions are equally easy in two dimensions, giving analytic expressions for the chemical potential. The reader may remember that both cases in three dimensions were discussed in Chapter 10 and that the algebra, particularly in the case of the gas of bosons, presented some awkwardness. We now show how the arguments proceed in two dimensions. The first step is to obtain an expression for the density of states. We remember from quantum mechanics that the allowed energies of a particle restricted to motion within a square of side L are given by

$$\epsilon = \frac{h^2}{8mL^2}(n_x^2 + n_y^2) \tag{12.116}$$

where n_x and n_y are non-zero positive integers. This equation is obviously very similar to (12.107) and the argument which follows is also similar. If the allowed states are represented on a graph of n_x versus n_y then they will appear as a square array of dots with unit spacing. Since (12.116) is the equation of a circle whose centre is at the origin and whose radius is $(8mL^2\epsilon/h^2)^{1/2}$, we can deduce that the number of states with energy less than ϵ is approximately equal to the 'area' of the positive quadrant of this circle, i.e.,

$$\tfrac{1}{4}\pi 8mL^2\epsilon/h^2 = \frac{2\pi mL^2\epsilon}{h^2} \tag{12.117}$$

Now if we differentiate this expression with respect to ϵ and multiply by a spin degeneracy factor g we obtain the required density of states in a form which is very accurate for large ϵ:

$$g(\epsilon) = \frac{2\pi gmL^2}{h^2} \tag{12.118}$$

Interestingly, this is independent of ϵ, whereas in the three-dimensional case $g(\epsilon)$ is proportional to $\epsilon^{1/2}$ which complicates the algebra. In order to find the chemical potential μ it is only necessary to use

$$N = \int_0^\infty g(\epsilon)\langle n(\epsilon)\rangle d\epsilon \tag{12.119}$$

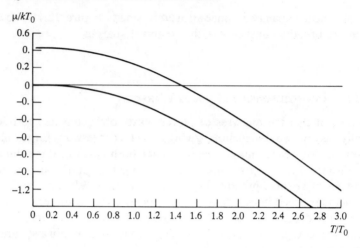

Figure 12.12 The chemical potentials of perfect two-dimensional Fermi and Bose gases, plotted from equations (12.121) as μ/kT_0 against T/T_0 where $T_0 \equiv Nh^2/2\pi mgkL^2$. The upper curve is for the Fermi gas and the lower for the Bose gas.

where, as usual,

$$\langle n(\epsilon)\rangle = \frac{1}{\exp((\epsilon-\mu)/kT) \pm 1} \qquad (12.120)$$

with the upper (+) sign being appropriate to fermions and the lower (−) to bosons. The integral can be done explicitly with either sign, and solving for μ we find

$$\mu = kT\ln\left[\pm\left\{\exp\left(\pm\frac{Nh^2}{2\pi mgkTL^2}\right) - 1\right\}\right] \qquad (12.121)$$

These functions are plotted in Figure 12.12. Armed with this knowledge of the chemical potential, one can go on to a consideration of adsorption isotherms (by equating chemical potentials in the adsorbed layer and in the bulk vapour as before) or to calculation of thermodynamic properties such as heat capacity. We shall not pursue this matter further here, but the principles are straightforward and the reader may wish to try the algebra.

It is worth mentioning that the idealized calculations presented above make assumptions about the nature of the adsorbed phase. The reality is much more interesting in that experiments over the last decade or so have revealed rich variety in the behaviour of adsorbed, 'restricted geometry' layers, including phase transitions of many kinds. Figure 12.13 shows an example of the adsorption isotherms of argon gas on a $MgBr_2$ substrate with clear evidence of a gas–liquid phase transition and a critical point.[9]

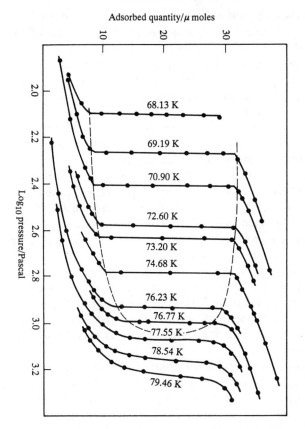

Figure 12.13 Adsorption isotherms of Ar on $MgBr_2$ at coverages up to a monolayer; figure adapted from F. Millot, Y. Lahrer, and C. Tessier, *J. Chem. Phys.* (1982), **76**, 3327.

There has also been work on a variety of other systems revealing transitions to solid phases.[10]

Finally, this chapter has certainly not exhausted the range of applications of the principle of equalization of chemical potentials, but we have tried to give an insight into the power of the method of giving a range of different cases. So far in this book we have concentrated on model systems in which interactions can be neglected. These in fact cover a surprising range of phenomena. The next chapter is devoted to a particular case where interactions can be taken into account.

CHAPTER 12: SUMMARY OF MAIN POINTS

1. The condition for the equilibrium between two phases is that the chemical potential of each phase should be equal.

2. Using the canonical distribution, the chemical potential of an Einstein solid can be calculated. If this is equated to the chemical potential of a classical gas, the vapour pressure of the solid can be calculated.
3. For an *ideal* liquid mixture, it is possible to find the chemical potentials of each of the two components. By equating each of these to their corresponding values in the gas phase, it is possible to derive Raoult's law of partial pressures.
4. By equating the chemical potentials of the different components of a mixture in two different phases, it is possible to calculate the equilibrium concentrations in the two different phases.
5. The theory is applied to the mixtures of liquid helium three and liquid helium four. The agreement with experiment is extremely good.
6. The ideal liquid mixture model can be used to calculate the osmotic pressure between two different concentrations separated by a semi-permeable membrane. The model predicts van't Hoff's formula for the osmotic pressure of weak solutions.
7. By treating the up spin electrons as one Fermi gas and the down spin electrons as another, it is possible, by equating their respective chemical potentials, to calculate the paramagnetic susceptibility.
8. If the chemical potentials of the electron gases of two different metals in contact are equated, the contact potential between the metals can be calculated.
9. It is possible to evaluate the canonical partition function of a model of multi-layer adsorption on a surface. By equating the chemical potentials of the adsorbate and the vapour, it is possible to derive the Brunauer–Emmett–Teller equation for multi-molecular adsorption.
10. The adsorbed layer can be approximated by a two-dimensional Debye solid.

QUESTIONS

12.1 Show that U and F can be written in forms

$$U = N\Phi_1\left(\frac{S}{N}, \frac{V}{N}\right)$$

$$F = N\Phi_2\left(\frac{V}{N}, T\right)$$

where Φ_1 and Φ_2 are functions appropriate to the system in question. Hence show that μ *cannot* be regarded as internal energy per particle or as Helmholtz free energy per particle. Only the Gibbs free energy will do.

12.2 At 115 K the vapour pressure of liquid argon (atomic weight 40) is 9.4×10^5 pascals ($\sim$ 9 atmospheres) and its density is 1.07×10^3 kg/m^3. At the same temperature the vapour pressure of liquid krypton (atomic weight 84) is 6.9×10^4 pascals ($\sim$ 0.7 atmospheres) and its density is 2.34×10^3 kg/m^3. Check on the validity of the approximation of equation (12.54) to Raoult's law, equation (12.55). Make a prediction for the vapour pressure of a liquid mixture of argon and krypton containing 30 % of argon atoms at 115 K.

12.3 Question 9.7 asked for a calculation of the variation of pressure with height in an isothermal atmosphere. Repeat the calculation, using the much simpler idea of equating the chemical potentials at heights 0 and h.

12.4 Consider a system in which an ideal gas of N molecules is introduced into a volume V where an adsorbent surface exists. Assume that there are $M > N$ equivalent sites in the surface where a gas molecule could be adsorbed, and where its energy relative to one at rest in the gas would be $-\epsilon_0$.

(a) Show that the grand partition function for the adsorbed molecules is given by $\Xi = (1 + \lambda_a q_a)^M$ where $\lambda_a = \exp(\mu_a/kT)$, μ_a is the chemical potential of the adsorbed phase, and $q_a = \exp(\epsilon_0/kT)$.

(b) Find μ_a and show that f, the ratio of the adsorbed molecules to sites, is given by $f = \lambda_a q_a / (1 + \lambda_a q_a)$.

(c) Find how f depends on the pressure in the gas phase.

12.5 In the previous question the adsorbed molecules were assumed to be attached to particular sites. This is not necessarily the case. Assume alternatively that the adsorbed molecules are free to move laterally and may then form a two-dimensional gas.

(a) Calculate the chemical potential, taking the energy of a molecule in the two-dimensional gas to be $p^2/2m - \epsilon_0$ where ϵ_0 is the binding energy.

(b) Calculate the number of adsorbed molecules in terms of the surrounding gas pressure p.

12.6 The minimum energy required to remove a conduction electron from a metal is $e\Phi = W - \epsilon_f$, where W is characteristic of the particular metal, ϵ_f is the Fermi energy inside the metal and Φ is called the work function. Consider an electron gas outside the metal in thermal equilibrium with the electrons in the metal at the temperature T. The density of electrons outside the metal is quite small at all laboratory temperatures where $kT \ll e\Phi$. By equating chemical potentials for the electrons outside and inside the metal, find the mean number of electrons per unit volume in the 'vapour' of electrons outside the metal.

12.7 Obtain an expression for the vapour pressure of a Debye solid with characteristic temperature θ_D whose atoms have, in the rest-equilibrium position, an energy $-\epsilon_0$ relative to an atom at rest in the vapour.

12.8 Verify that a 1 % ^{3}He impurity in liquid ^{4}He at 1.5 K satisfies the classical criterion $T \gg T_f$ and use van't Hoff's formula (12.76) to evaluate the osmotic pressure of the ^{3}He relative to pure liquid ^{4}He (atomic weight 4) whose density is 145 kg/m^3. Express the pressure in pascals and in terms of the height of a liquid helium column.

12.9 Prove Curie's law, equation (12.86), for a classical assembly of magnetic moments.

12.10 A gas of N' weakly interacting particles, each of mass m, adsorbed on a surface of area A on which they are free to move, can form a two-dimensional ideal gas on such a surface. The energy of an adsorbed particle is

$$\left[\frac{h^2}{8m} \frac{(r^2 + s^2)}{A} - \epsilon_0 \right]$$

where r and s are positive integers and ϵ_0 is the binding energy which holds the particle on the surface.

(a) Show that the canonical partition function Q can be written in the form

$$Q = \frac{1}{N!} \left(\frac{2\pi m A k T}{h^2} \exp\left(\frac{\epsilon_0}{kT} \right) \right)^N$$

Deduce from it an expression for the chemical potential μ' of this adsorbed ideal gas. You may safely replace sums by integrals so as to obtain a convenient form for Q and you may quote the result

$$\int_0^\infty \exp(-x^2)\, dx = \frac{\sqrt{\pi}}{2}$$

(b) At temperature T the equilibrium condition between particles adsorbed on the surface and in the surrounding three-dimensional gas can be written in terms of the respective chemical potentials. Use this condition to show that the mean number n' of particles adsorbed per unit area of surface when the mean pressure of the surrounding gas is p is

$$n' = \frac{p}{(kT)^{\frac{3}{2}}} \left(\frac{h^2}{2\pi m}\right)^{\frac{1}{2}} \exp\left(\frac{\epsilon_0}{kT}\right)$$

The chemical potential μ of a three-dimensional ideal gas containing N/V particles per unit volume may be assumed to be

$$\mu = -kT \ln\left[\frac{V}{N}\left(\frac{2\pi mkT}{h^2}\right)^{\frac{3}{2}}\right]$$

REFERENCES

1. Rowlinson J.S. and Swinton F.L., *Liquids and Liquid Mixtures*, 3rd ed. (Butterworths, 1982).
2. Edwards D.O. and Daunt J.G., *Phys. Rev.* (1961), **124**, 640.
3. Edwards D.O., Brewer, D.F., Seligman P., Skertic M.M. and Yaqub M., *Phys. Rev. Lett.* (1965), **15**, 773.
4. Ifft E.M., Edwards D.O., Sarwinski R.E. and Skertic M.M., *Phys. Rev. Lett.* (1967), **19**, 831; Edwards D.O., Ifft E.M., Sarwinski R.E., *Phys. Rev.* (1969), **177**, 380.
5. Wansink D.H.N. and Taconis K.W., *Physica* (1957), **23**, 125.
6. Bleaney B.I. and Bleaney B., *Electricity and Magnetism*, 3rd ed. (Oxford, 1976).
7. Husa D.L., Edwards D.O. and Gaines J.R., *Phys. Lett.* (1966), **21**, 28; Anderson A.C., Edwards D.O., Roach W.R., Sarwinski R.E. and Wheatley J.C., *Phys. Rev. Lett.* (1966), **17**, 367.
8. Anderson P., *Phys. Rev.* (1952), **88**, 655.
9. For example, Millot F., Lahrer Y. and Tessier C., *J. Chem. Phys.* (1982), **76**, 3327.
10. For example, Thomy A. and Duval X., *Journal de Chimie Physique* (1970), **67**, 1101.

CHAPTER 13

The Interacting Classical Gas

13.1 Classical statistical mechanics
13.2 The interacting gas
13.3 Evaluation of the second virial coefficient
Questions

13.1 Classical statistical mechanics

The whole development of the methods of statistical mechanics in this book has been largely based on our knowledge that energies in systems are often quantized. We have further avoided complexity by treating systems as if they had no potential energies of interaction. This is certainly true in perfect gases, and in solids we were able to identify sub-systems (oscillators or modes), which could be treated *as if* they were non-interacting. In liquids this is generally impossible (except for the liquid isotopes of helium whose 'excitations', for example photons and rotons in liquid ^{4}He-II and quasi-particles in liquid ^{3}He, are treated as non-interacting). In a sense, we have been fortunate that mathematical techniques are available which enabled us to make as much progress as we have. In this chapter we introduce a technique which, for gases and liquids, makes it possible to include the effects of interaction between atoms. Our limited objective is to derive an equation of state which contains first-order correction terms due to interactions. We shall in fact derive the van der Waals equation of state. The approach is suggestive rather than strictly rigorous, since in a short chapter we cannot hope to follow up every algebraic question mark which may arise. However, we feel that even an undergraduate introduction to statistical

CLASSICAL STATISTICAL MECHANICS 231

mechanics must give some clue as to how the methods can be applied in the dirty world of real materials.

No results are obtainable from classical (i.e., non-quantized) statistics which cannot be found in principle as limiting cases of quantized statistics but often the classical version is easier (interacting gases are a case in point). What, then, is the classical version of the canonical partition function? A clue may be found by looking at two cases.

For a single one-dimensional simple harmonic oscillator, equation (5.15) gives

$$q = \tfrac{1}{2}\operatorname{cosech}\left(\frac{h\nu}{2kT}\right) \qquad (13.1)$$

At high temperatures $(T \gg h\nu/2k)$, $q \to kT/h\nu$, and we wish to know whether this latter result could have been obtained without assuming details of the quantum mechanical SHO. The so-called correspondence principle has a bearing here in that high temperatures bring high quantum numbers into play. At high enough temperatures we would expect the discreteness of the energy levels to be unimportant, which would seem to imply that we could then always replace sums by appropriate integrals. We have already met examples of this, for instance in the classical gas (Section 9.1) and in the rotation of diatomic molecules (Section 5.4). In the latter case we approximated equation (5.24) by an integral form

$$q = \sum_{J=0}^{\infty} (2J+1)\exp\left(-\frac{\hbar^2 J(J+1)}{2IkT}\right)$$

$$\to \int_0^{\infty} (2J+1)\exp\left(-\frac{\hbar^2 J(J+1)}{2IkT}\right) dJ$$

$$= \frac{T}{\theta_r}$$

where

$$\theta_r \equiv \frac{\hbar^2}{2Ik}$$

Now we could have argued that for high J-values which are important at high T (i.e., $T \gg \theta_r$) the angular momentum is approximately $L = \hbar J$ so that the integral representation is

$$q = \frac{1}{h^2}\int_0^{\infty} 8\pi^2 \exp\left(-\frac{L^2}{2IkT}\right) L\,dL \qquad (13.2)$$

The appearance of this equation suggests that the integral need not contain h which only occurs in a multiplicative constant. Actually the rigid rotator is properly regarded as a system with two degrees of freedom. A simpler case is the simple harmonic oscillator (SHO).

Classically, the energy of an SHO is $E = 2\pi^2 m\nu^2 x^2 + p^2/2m$. The classical 'state' is specified by x and p, both of which may take values in the continuous range $-\infty$ to $+\infty$. It is plausible to expect an integral in place of a sum so we try

$$q_{class} = c \int_{-\infty}^{\infty} \int_{-\infty}^{\infty} \exp\left(-\frac{E}{kT}\right) dx dp$$

$$= c \int_{-\infty}^{\infty} \exp\left(-\frac{2\pi^2 m\nu^2 x^2}{kT}\right) dx \int_{-\infty}^{\infty} \exp\left(-\frac{p^2}{2mkT}\right) dp \quad (13.3)$$

where c is a constant to be determined. Integrating,

$$q_{class} = c \left(\frac{\pi kT}{2\pi^2 m\nu^2}\right)^{1/2} (2\pi mkT)^{1/2} \quad (13.4)$$

$$= c \frac{kT}{\nu} \quad (13.5)$$

This suggests that c be chosen to equal $1/h$ so that equation (13.5) should agree with the high-temperature limit of (13.1). Hence

$$q_{class} = \frac{1}{h} \int_{-\infty}^{\infty} \int_{-\infty}^{\infty} \exp\left(-\frac{E}{kT}\right) dx dp \quad (13.6)$$

Here only $1/h$ appears (rather than $1/h^2$ as in equation (13.2)) since we are dealing with a one-dimensional case.

For a three-dimensional example consider the particle in a box. We found in Section 5.5 that the summation was evaluated by conversion to an integral, a process which was found to be appropriate, when $T \gg \theta_t$. Then

$$q = V \left(\frac{2\pi mkT}{h^2}\right)^{3/2} \quad (13.7)$$

Classically $E = (p_x^2 + p_y^2 + p_z^2)/2m$ and the spatial coordinates do not appear because the energy is taken to be purely kinetic. By analogy with equation (13.6) we try

$$q_{class} = \frac{1}{h^3} \iiint\iiint \exp\left[-\frac{p_x^2 + p_y^2 + p_z^2}{2mkT}\right] dp_x dp_y dp_z dx dy dz \quad (13.8)$$

where the momentum limits run from $-\infty$ to $+\infty$ and the spatial limits span the dimensions of the box. In this case, where E has no spatial dependence,

$$\iiint dx\,dy\,dz = V \tag{13.9}$$

where V is the volume of the box and the remaining factor in (13.8) is an integral over momenta which can be explicitly integrated. Hence

$$q_{class} = V\left(\frac{2\pi mkT}{h^2}\right)^{3/2} \tag{13.10}$$

which agrees exactly with (13.7). Incidentally, the inclusion of $1/h^3$ in this case is intuitive because we are dealing with a three-dimensional sub-system. By extension one would expect that for a system of N indistinguishable sub-systems,

$$Q_{class} = \frac{1}{N!h^{3N}} \iiint \cdots \iiint \exp\left(-\frac{E}{kT}\right) dp_{x_1}\,dp_{y_1}\,dp_{z_1}\,dp_{x_2} \cdots$$
$$dx_1\,dy_1\,dz_1\,dx_2 \cdots dz_N \tag{13.11}$$

where the integral is a $6N$-fold multiple integral covering momentum and spatial coordinates for each of the N particles.

Expression (13.11) is called a phase integral. The reader was in fact led to other examples of phase integrals in Questions 9.6 and 9.7. We are now in a position to use the topic of Question 9.7 in particular as an illustrative case here before proceeding further. Consider a Maxwell–Boltzmann gas contained in a cube of volume $V = L^3$ which is in a uniform gravitational field g directed along the x-axis. Then

$$E = \sum_{i=1}^{N} \frac{p_{x_i}^2 + p_{y_i}^2 + p_{z_i}^2}{2m} + mg\sum_{i=1}^{N} x_i \tag{13.12}$$

The second term, representing the gravitational potential energy, takes a particularly simple form here so that the phase integral (13.11) can be exactly evaluated by factorizing it:

$$Q_{class} = \frac{1}{N!h^{3N}} \prod_{i=1}^{N} I_i \tag{13.13}$$

where

$$I_i = \iiint_{-\infty}^{+\infty} \exp\left[-\frac{p_{x_i}^2 + p_{y_i}^2 + p_{z_i}^2}{2mkT}\right] dp_{x_i} dp_{y_i} dp_{z_i}$$

$$\times \iiint_0^L \exp\left[-\frac{mgx_i}{kT}\right] dx_i dy_i dz_i \qquad (13.14)$$

$$= (2\pi mkT)^{3/2} \frac{kTV}{mgL}\left[1 - \exp\left(-\frac{mgL}{kT}\right)\right] \qquad (13.15)$$

Thus

$$Q_{class} = \left(\frac{2\pi mkT}{h^2}\right)^{3N/2} \left(\frac{eV}{N}\right)^N \left(\frac{kT}{mgL}\right)^N \left[1 - \exp\left(-\frac{mgL}{kT}\right)\right]^N \qquad (13.16)$$

Note that g in these equations is the acceleration due to gravity, and *not* the spin degeneracy. Expression (13.16) reduces to equation (9.15) as $g \to 0$ as we should expect.

It is important to remember that the use of equation (13.11) is restricted to 'high temperatures', but the form of the equation gives no clue as to a definition of 'high'. For that we have to be sure that $kT \gg \Delta$ where Δ is a characteristic energy level spacing. For gases we know that all is well in that respect since, as we have said elsewhere, Δ/k is generally far below any physically accessible temperature. We shall now assume that equation (13.11) is the correct form for the partition function in the classical limit and use it to find the effects of atomic interactions on the properties of a gas.

13.2 The interacting gas

We propose to employ equation (13.11), taking

$$E = \frac{1}{2m}(p_{x_1}^2 + \ldots + p_{z_N}^2) + \Phi(x_1 \ldots z_N) \qquad (13.17)$$

The momentum integrations can be carried out immediately:

$$Q_{class} = \left(\frac{2\pi mkT}{h^2}\right)^{3N/2} \left(\frac{e}{N}\right)^N \int \ldots \int \exp\left[-\frac{\Phi(x_1 \ldots z_N)}{kT}\right] dx_1 \ldots dz_N \qquad (13.18)$$

$$= \left(\frac{2\pi mkT}{h^2}\right)^{3/2N} \left(\frac{e}{N}\right)^N Z_N \qquad (13.19)$$

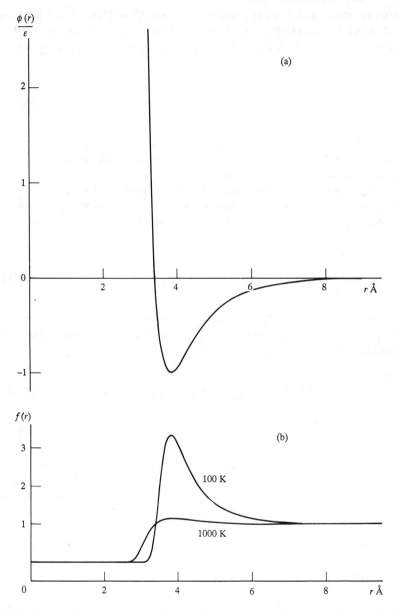

Figure 13.1 (a) The interaction energy $\varphi(r)/\epsilon$ between two argon atoms as given by equation (1.1). (b) The corresponding quantity $f(r)$ as defined by equation (13.21) for two different temperatures. For most values of r, $f(r)$ is either 1 or 0 and this is particularly true at high temperatures.

where we have used Stirling's approximation $N! \simeq (N/e)^N$. Z_N is called the classical configuration integral and, if $\Phi = 0$, $Z_N = V^N$ as expected. If the interaction is such that a pair of atoms have interaction energy $\varphi(r_{ij})$, then

$$\Phi = \sum_{i>j} \varphi(r_{ij}) \tag{13.20}$$

where r_{ij} is the distance between atoms i and j. The restriction $i > j$ is merely to avoid counting energies twice in covering all possible pairs. A typical shape for $\varphi(r)$ is shown in Figure 13.1(a).

The evaluation of the configuration integral cannot be carried out exactly, and so an approximation procedure has to be used. We give a procedure first derived by van Kampen,[1] in which we start by defining the function

$$f_{ij} = \exp\left[-\frac{\varphi(r_{ij})}{kT}\right] \tag{13.21}$$

and noting from Figure 13.1(b) that f_{ij} is nearly everywhere unity except when the molecules i and j are close together. In terms of the f_{ij} the configuration integral is given by

$$Z_N = \int \ldots \int f_{12} f_{13} \ldots f_{23} f_{24} \ldots f_{(N-1)N}\, dx_1 \ldots dz_N \tag{13.22}$$

$$= V^N \overline{f_{12} f_{13} \ldots f_{23} f_{24} \ldots f_{(N-1)N}} \tag{13.23}$$

where the bar denotes an average of the product over the variables $x_1 \ldots z_N$, i.e., it is the integral in equation (13.22) divided by V^N. Now if we had just the product of two of the fs, say $f_{12} f_{13}$, then since each f depends only on the difference of two coordinates we could write

$$\overline{f_{12} f_{13}} = \overline{f_{12}}\, \overline{f_{13}} = \left(\overline{f_{12}}\right)^2 \tag{13.24}$$

For the triplet product one cannot strictly write

$$\overline{f_{12} f_{13} f_{23}} = \left(\overline{f_{12}}\right)^3 \tag{13.25}$$

However, for low densities equation (13.25) may be a good approximation since if any one of the fs were equal to one then we could apply equation (13.24) to the remaining two. Furthermore, the configurations when all three fs differ from unity is very rare if the density is low. We can extend this argument to the product of N of the functions and hence write as a first approximation to equation (13.23),

$$Z_N^{(1)} = V^N \left(\overline{f_{12} f_{13}} \cdots \overline{f_{23} f_{24}} \cdots \overline{f_{(N-1)N}} \right) = V^N \left(\overline{f_{12}} \right)^{N(N-1)/2} \quad (13.26)$$

In the thermodynamic limit $N \to \infty$, $V \to \infty$, $N/V = \rho$ is finite, we have

$$\text{Lt } Z_N^{(1)} = \text{Lt } V^N \left(\frac{1}{V^2} \int \cdots \int \exp\left[-\frac{\varphi(r_{12})}{kT} \right] dx_1 \cdots dz_2 \right)^{N(N-1)/2} \quad (13.27)$$

$$= \text{Lt } V^N \left[\left(1 + \frac{\rho}{N} \iiint \left\{ \exp\left[-\frac{\varphi(r)}{kT} \right] - 1 \right\} dxdydz \right)^N \right]^{N/2} \quad (13.28)$$

$$= V^N \exp\left[-\frac{\rho N B(T)}{2} \right] \quad (13.29)$$

where

$$B(T) = -\iiint \left\{ \exp\left[-\frac{\varphi(r)}{kT} \right] - 1 \right\} dxdydz$$

$$= -4\pi \int \left\{ \exp\left[-\frac{\varphi(r)}{kT} \right] - 1 \right\} r^2 dr \quad (13.30)$$

In passing from equation (13.29) to equation (13.30) we have transformed from integrals over the centre of mass coordinate $R = \frac{1}{2}(r_1 + r_2)$ and relative coordinates $r = r_1 - r_2$. We have also used the binomial expansion, the fact that $\iiint dxdydz = V$, and the approximation $N - 1 = N$.

To get the second approximation we note that our first approximation treated the product of two fs correctly but failed when it came to average products of the form $\overline{f_{12} f_{13} f_{23}}$. Now in our product of N fs there will be $^N C_3$, i.e., $N(N-1)(N-2)/3$ products of this form, so to correct equation (13.29) we must multiply it by a factor

$$\left(\frac{\overline{f_{12} f_{13} f_{23}}}{\overline{f_{12}} \, \overline{f_{13}} \, \overline{f_{23}}} \right)^{N(N-1)(N-2)} \quad (13.31)$$

We leave it as an exercise for the reader to show that in the thermodynamic limit this leads to the result

$$Z_N^{(2)} = V^N \exp\left[-\frac{\rho N B(T)}{2} + \frac{\rho^2 N^2 C(T)}{3} \right] \quad (13.32)$$

where

$$C(T) = \frac{1}{2}\overline{f_{12}f_{13}f_{23}} = \frac{1}{2V^3}\int \ldots \int f_{12}f_{13}f_{23}\,dx_1 \ldots dz_3 \quad (13.33)$$

For the general term in this expansion the reader is referred to the original paper of van Kampen.

For the rest of this discussion we shall use the first approximation (13.29) involving the so-called second virial coefficient $B(T)$. Substituting equation (13.29) into equation (13.19) and taking the logarithm we find

$$F = -kT \ln Q_{class}$$
$$= -NkT\left\{\frac{3}{2}\ln\left[\frac{2\pi mkT}{\hbar^2}\right] + \ln\left[\frac{eV}{N}\right] - \frac{\rho NB(T)}{2}\right\} \quad (13.34)$$

All the thermodynamic properties can now be calculated from equation (13.34); most of these are left as an exercise for the reader (see Question 13.3)). We shall deal here only with the equation of state. Now

$$p = kT\left(\frac{\partial \ln Q}{\partial V}\right)_{T,N}$$
$$= \frac{NkT}{V}\left[1 + \frac{N}{V}B(T)\right] \quad (13.35)$$

Remembering that we have taken only correction terms proportional to N/V, we may regard equation (13.35) as the beginning of a series expansion for the pressure in the form

$$p = \frac{NkT}{V}\left[1 + \frac{N}{V}B(T) + \left(\frac{N}{V}\right)^2 C(T)\ldots\right] \quad (13.36)$$

This equation whose first three terms can be derived from $Z_N^{(2)}$, equation (13.32), is called a 'virial expansion', with $B(T)$, $C(T)$ ... called the second, third, etc., virial coefficients. An alternative expansion in terms of pressures rather than number densities easily follows but we shall not discuss that here.

13.3 Evaluation of the second virial coefficient

Figure 13.2 shows the actual function $B(T)$ obtained for argon from detailed measurements of p, V and T. The second virial coefficient can be obtained by numerical integration of equation (13.30) using a known or assumed $\varphi(r)$. Such a calculation has been tabulated[2] and the agreement for argon (whose interaction is well-known, see equation (1.1)) is seen in Figure 13.2 to be good.

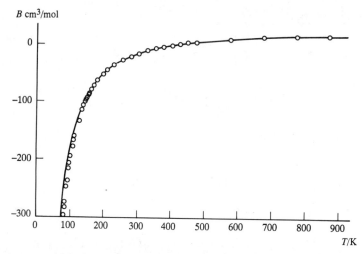

Figure 13.2 The second virial coefficient for argon obtained from experiments by R.D. Weir, I. Wynn Jones, J.S. Rowlinson and G. Saville, *Trans. Faraday Soc.* (1967), **63**, 1320, A. Michels, J.M. Levelt and W. de Graaff, *Physica* (1958), **24**, 659 and E. Whalley, Y. Lupien and W.G. Schneider, *Canad. J. Chem.* (1953), **31**, 722. The continuous curve is from the theoretical tabulation by J.O. Hirschfelder, C.F. Curtiss and R.B. Bird in *Molecular Theory of Gases and Liquids* (Wiley, 1965) appropriate to the interaction of equation (1.1). (It is interesting to observe that although data and theory agree above about 130 K there is a noticeable systematic discrepancy below that temperature which suggests that equation (1.1) should be modified. Figure 13.1(b) show that $f(r)$ is more sensitive at lower temperatures to the details of the minimum in $\varphi(r)$, and also specifically quantum-mechanical scattering effects become more important at low temperatures.) For argon, the Boyle temperature, at which $B = 0$, is 410 K.

However there is some value in obtaining an approximation which reveals the physics without being exact. Such an approximation is the van der Waals equation of state, and it can be derived as follows. It is true of most interatomic interactions that there is a strong repulsion at separations below a certain value. We may conveniently define a 'hard core' diameter d as the value of r for which $\varphi(r) = kT$. This quantity d is thus strictly a function of temperature but not (see Question 13.3) a strong function because the interaction energy (see Figure 13.1(a)) rises so sharply with decreasing separation. We may treat d as a constant without serious error. Then the integral in (13.30) breaks naturally into two parts because

$$\left. \begin{aligned} \exp\left[-\frac{\varphi(r)}{kT}\right] - 1 &\approx -1 & \text{for } r \leq d \\ \exp\left[-\frac{\varphi(r)}{kT}\right] - 1 &\approx -\frac{\varphi(r)}{kT} & \text{for } r > d \end{aligned} \right\} \tag{13.37}$$

Hence, from equation (13.30),

$$B(T) = -\int_0^\infty \left\{ \exp\left[-\frac{\varphi(r)}{kT}\right] - 1 \right\} 2\pi r^2 \, dr$$

$$\approx 2\pi \int_0^d r^2 \, dr + \frac{2\pi}{kT} \int_d^\infty \varphi(r) r^2 \, dr$$

$$= b - \frac{a}{kT} \tag{13.38}$$

thus defining a and b. Hence, from equation (13.34),

$$F = -kT \ln Q$$

$$= -\frac{3}{2} NkT \ln\left[\frac{2\pi mkT}{h^2}\right] - NkT \ln\left[\frac{eV}{N}\right] - \frac{N^2 kT}{V}\left[\frac{a}{kT} - b\right] \tag{13.39}$$

The problem is now solved. We find, by the appropriate operations on F,

$$C_V = \frac{3}{2} Nk \tag{13.40}$$

as for the non-interacting gas, but

$$p = \frac{NkT}{V} - \frac{N^2 kT}{V^2}\left(\frac{a}{kT} - b\right) \tag{13.41}$$

or, on rearrangement,

$$\left[p + \left(\frac{N}{V}\right)^2 a\right] = \frac{NkT}{V}\left[1 + \left(\frac{N}{V}\right) b\right] \tag{13.42}$$

This expression could be left as it stands, but for the satisfaction of obtaining the familiar van der Waals equation of state we make the further approximation

$$\left[1 + \left(\frac{N}{V}\right) b\right] \approx \left[1 - \left(\frac{N}{V}\right) b\right]^{-1} \tag{13.43}$$

Thus

$$\left[p + \left(\frac{N}{V}\right)^2 a\right]\left[1 - \left(\frac{N}{V}\right) b\right] = \frac{NkT}{V} \tag{13.44}$$

where

$$a = -2\pi \int_0^\infty \varphi(r) r^2 dr \qquad (13.45)$$

and

$$b = \frac{2}{3}\pi d \qquad (13.46)$$

This is as far as we propose to pursue this matter of non-ideality. More detailed treatments can be found elsewhere.[1]

CHAPTER 13: SUMMARY OF MAIN POINTS

1. Classical statistical mechanics historically preceded quantum mechanics but no results are obtainable classically which cannot be obtained as limiting cases of quantized statistics, integrals replacing sums.
2. Indication of how a classical approach can be taken towards the ideal non-degenerate gas.
3. The point of taking this view is that the inclusion of realistic interactions between the atoms is perhaps easier in the classical approach. This is demonstrated by adopting a typical form for the atomic interaction energy and using it to obtain the second virial coefficient which is shown to compare well with measurements.
4. Further, by considering the equation of state including the second virial coefficient it is shown that the familiar van der Waals equation of state can be derived as a good approximation.

QUESTIONS

13.1 Complete the algebra leading from equation (13.31) to equation (13.32).

13.2 Using equation (13.34) for F, derive the thermodynamic functions S, U, F, C_V, and μ.

13.3 For the argon interaction (1.1) find the hard-core diameter $d(T)$ for temperatures $T = 150$ K, 300 K, 1000 K, and 3000 K to show that it is a weak function of T. Note that this is a justification of equation (13.37).

242 THE INTERACTING CLASSICAL GAS

13.4 An adsorbed surface layer of area A consists of N atoms which are free to move over the surface and can be treated as a classical two-dimensional gas. The atoms interact with each other according to a potential $\varphi(r)$ which depends only on their mutual separation r. Find the film pressure, i.e., the mean force per unit length, of this gas (up to terms involving the second virial coefficient).

13.5 The second virial coefficient $B(T)$ has been measured as a function of temperature for nitrogen gas and is given below. Assume that the intermolecular potential $\varphi(r)$ between the molecules is of the form shown in the diagram. From the data given, determine what you consider to be the best choices for the constants σ, r_0 and ϵ.

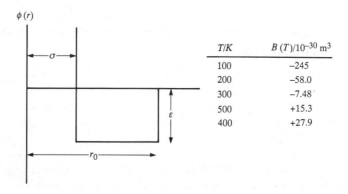

T/K	$B(T)/10^{-30}$ m^3
100	−245
200	−58.0
300	−7.48
500	+15.3
400	+27.9

13.6 Adopt an intermolecular potential of the shape given in problem (13.5) and use it to compute the first correction dependent on the particle density N/V to the perfect-gas expression for μ in terms σ, r_0 and ϵ.

13.7 Each molecule of a classical gas has a permanent electric dipole. The gas is subject to an electric field E.

Show that the polarization is proportional to the field for sufficiently small $|E|$ and that the next approximation is

$$P = \alpha E + \gamma |E|^2 E$$

Find the value of γ and explain physically why there is no term proportional to E^2 and why γ is negative.

13.8 Sketch a plot of the function $f(r) = \exp[-\varphi(r)/kT]$ for hard spheres and for the Lennard-Jones potential (at several different temperatures) $\varphi(r) = 4\epsilon[(\sigma/r)^{12} - (\sigma/r)^6]$ where, for xenon, $\epsilon = 3.2 \times 10^{-21}$ J and $\sigma = 0.398$ nm.

Find the second virial coefficient at high temperatures if an attractive tail of the form $-\epsilon(d/r)^6$ is added to the hard sphere potential.

13.9 Using the anharmonic potential $U(x) = cx^2 - gx^3 - fx^4$, show that the approximate heat capacity of the classical anharmonic oscillator in one dimension is

$$k\left[1 + \left\{\left(\frac{3f}{2c^2}\right) + \left(\frac{15g^2}{8c^2}\right)\right\}kT\right]$$

Also, using the xenon interaction energy given in Question 13.8, evaluate c, g, and f, and hence estimate the relative importance of the anharmonic terms. (Such anharmonic effects were referred to in Section 9.3 relating to diatomic gases.) Also estimate the mean value $\langle x \rangle$ and show that it increases with temperature.

13.10 In the last question we referred to a classical anharmonic oscillator. The energy levels of a *quantum-mechanical* anharmonic oscillator are approximately

$$E_n = (n + \tfrac{1}{2})\hbar\omega - \alpha(n + \tfrac{1}{2})^2\hbar\omega$$

where α is the (small) degree of anharmonicity. Show that to first order in α and fourth order in $\theta_v/T (\equiv \hbar\omega/kT)$, the heat capacity of N such oscillators is

$$C = Nk\left[\left\{1 - \frac{1}{12}\left(\frac{\theta_v}{T}\right)^2 + \frac{1}{240}\left(\frac{\theta_v}{T}\right)^4\right\} + 4\alpha\left\{\left(\frac{T}{\theta_v}\right) + \frac{1}{80}\left(\frac{\theta_v}{T}\right)^3\right\}\right]$$

13.11 For a real low-density gas obtain the internal energy up to terms proportional to its density in terms of the intermolecular potential $\varphi(r)$.

The radial distribution function $g(r, \rho, T)$ is defined as follows. Consider a gas at temperature T in a volume V and let $\rho = N/V$ be its mean particle density. Then

$$\rho g(r, \rho, T) 4\pi r^2 \, dr$$

is the number of molecules which lie within a spherical shell of radius r and thickness dr, given that there is a molecule at the origin $r = 0$. For a uniform distribution we would have $g \equiv 1$. The g allows for deviations from uniform density due to the intermolecular forces. (Experimentally one obtains g from X-ray diffraction experiments.) Show that for a real gas the internal energy is related to the radial distribution function by the equation

$$E = \frac{3}{2}NkT + \frac{1}{2}N\rho \int_0^\infty 4\pi r^2 g(r, \rho, T)\varphi(r) \, dr$$

244 THE INTERACTING CLASSICAL GAS

Obtain a relation between the radial distribution function in the limit of zero density and the intermolecular potential $\varphi(r)$.

13.12 Show that the entropy of an interacting monatomic gas can be written in the approximate form

$$S(T, N/V) = S_{ideal}(T, N/V) + A(T) N/V$$

where $S_{ideal}(T, N/V)$ corresponds to $\varphi(r) = 0$. Obtain an expression for $A(T)$ in terms of $\varphi(r)$ and show that it is always negative. This is an illustration of the statement that correlations always decrease entropy.

13.13 Consider a one-dimensional gas of N rods of length d, moving on a line segment of length L. Calculate the configuration integral and show that in the thermodynamic limit ($N \to \infty$, $L \to \infty$, N/L finite) the equation of state gives the force (analogue of pressure) as

$$f = \frac{NkT}{L - Nd}$$

REFERENCES

1 van Kampen N.G., *Physica* (1961), **27**, 783.
2 Hirschfelder J.O., Curtiss C.F. and Bird R.B., *The Molecular Theory of Gases and Liquids* (Wiley, 1965).

CHAPTER 14

Modern Approaches to Phase Transitions

14.1 Introduction
14.2 The Ising model
14.3 The mean field approximation
14.4 The renormalization group and the one-dimensional Ising model
14.5 The renormalization group and the two-dimensional Ising model
14.6 Monte Carlo methods
14.7 Phase changes and dimensionality
Questions

14.1 Introduction

In this final chapter we return to the subject of magnetic systems, though, as we shall see, we will be able to generalize to other types of system. In Chapter 7 we considered the example of a dilute paramagnetic salt where the magnetic atoms are very far apart and consequently the interactions between the spins can be ignored. This simplifying feature, plus the fact that the partition function for the individual spins was easily evaluated, enabled us to obtain an exact result. In this chapter we will relax the first assumption i.e., we will consider a limited form of interaction between the spins but continue to exploit the simplicity of their individual properties. Apart from the fact that in nature there are few systems where it it reasonable to ignore the interactions, it also turns out that including the interactions not only enables us to consider more realistic systems, but also to investigate a whole new phenomenon namely that of phase transitions. In the 1970s there was an explosion of interest in this subject, following the pioneering work of K.G. Wilson, who was awarded a Nobel prize for his efforts. Until that time efforts to understand phase transitions had been restricted to trying to obtain appropriate expansions on either side of the transition. Wilson's contribution was to recognize the relevance of the so-called renormalization group to this phenomenon.

We consider a simple model which takes into account some of the effects of interactions. The model, known as the Ising model, assumes an interaction between one component (usually taken as the z-component) of the spins of nearest neighbours. Such a model would correspond most closely to a crystal in which there existed a strong anisotropy, but the importance of the model goes beyond its application to magnetic systems. The reason for this is that in one and in two dimensions it is possible to evaluate the partition function exactly (this is not easy even in one dimension and the solution in two dimensions given by Onsager is a mathematical *tour de force*). Hence it is possible to evaluate all of the thermodynamic functions. In one dimension nothing of particular significance occurs, but in two dimensions the model exhibits a phase transition. In particular, near the critical temperature T_c and at zero external magnetic field, the specific heat is described by

$$\frac{C}{Nk} \propto T^{-2} \ln |(T - T_c)^{-1}| \tag{14.1}$$

and the magnetic moment by

$$M \propto (T_c - T)^{1/8} \quad \text{for } T < T_c \tag{14.2}$$

where the critical temperature is given by

$$T_c = 2.269 J/k \tag{14.3}$$

where J is the coupling constant between the nearest-neighbour spins (see Section 14.2 below). In three dimensions no analytic solution has been found and so approximate methods have to be resorted to. However, with the exact solutions known in one and two dimensions, it is possible to check the accuracy of the method, at least in those dimensions. Furthermore, the Ising model is isomorphic with other simple models, in particular the so-called lattice gas, which has been used as a model for liquid–gas phase transitions.

14.2 The Ising model

The energy levels of the Ising model can be expressed as

$$E_{\{s_i\}} = -\gamma B \sum_i s_i - \tfrac{1}{2} \sum_i \sum_{j \, NN} J s_i s_j \tag{14.4}$$

where $\{s_i\}$ represents the set of N quantum numbers that characterize the energy level and the subscript NN indicates that the sum over j is to be carried out over all the z nearest neighbours of the spin labelled i. Note the factor of $\frac{1}{2}$ which ensures that pairs of spins are not counted twice. From now on we consider a regular lattice with a spin on every site, so that z is a well-defined number. The first term on the right-hand side gives the energy levels of the non-interacting case given by (7.20), with degeneracies given by (7.21). The second term gives the effect of the interactions between the z components of nearest-neighbour interactions.

The task we have is to evaluate the canonical partition function for a system whose energy levels are given by equation (14.4). As we mentioned above, the exact solutions are difficult in one and two dimensions and have yet to be obtained in three; we have therefore to resort to approximate methods. The simplest is the mean field approximation (also sometimes known as the molecular field approximation).

14.3 The mean field approximation

Examination of the energy levels (14.4) shows that we could write them in the form

$$E_{\{s_i\}} = -\gamma \sum_i [B + \tfrac{1}{2} B_{int}(i)] s_i \tag{14.5}$$

where

$$B_{int}(i) = J \sum_{j\,NN} s_j / \gamma \tag{14.6}$$

is the internal field seen by the ith spin due to its z nearest neighbours. (The factor $\frac{1}{2}$ appears because we must not count the interactions between *pairs* twice.) Thus we can interpret the effect of the interactions as adding to the external field B, a local internal field acting on the spin due to immediate neighbours. This field will differ from spin to spin, depending upon the polarizations of the neighbours. One possible approximation would be to ignore the variation of the internal field and assume that each spin sees the same average field. This approximation is analogous to the Einstein model of a crystal, where each atom is assumed to vibrate in a potential well derived from putting its nearest neighbours in their average positions. In this case, however, there is one very important difference: for the Einstein solid, the potential well is assumed to be temperature-independent, since we can ignore the small increase between the average spacing of the atoms as the temperature increases. As we shall see, the analogous situation does not hold in the magnetic case. The mean field approximation consists of replacing the s_j in the expression for the internal field by its mean value. The mean value of s_j is simply the total macroscopic value of the spin (M/γ) divided by the total number of spins, i.e.,

$$\langle s_j \rangle = M/N\gamma = m/\gamma \tag{14.7}$$

Each spin therefore sees the same field which is the sum of the external field and the average field

$$B_{av} = mzJ/\gamma^2 \tag{14.8}$$

It is now very easy to calculate the magnetization, since we are dealing with independent spins and we can immediately use (7.26) and (7.23). The result is an equation for the magnetic moment per spin, i.e.,

$$m = \gamma \tanh\left[(\gamma B + mzJ/\gamma)/kT\right] \tag{14.9}$$

If $J = 0$ this reduces to our previous result. However, if the external field is zero we get the following equation for the spontaneous magnetization per particle:

$$m = \gamma \tanh\left[mzJ/\gamma kT\right] \tag{14.10}$$

It is not difficult to see that this has non-zero solutions provided that $zJ/kT > 1$. (If the reader sketches the right-hand side of equation (14.10) as a function of m and then draws the 45° line, it will be seen that the curves only intersect at $m \neq 0$ if the inequality does not hold.) So the critical temperature is predicted to be

$$T_c = zJ/k \tag{14.11}$$

For a square lattice in two dimensions, this gives $T_c = 4J/k$ compared to the exact result of $2.269J/k$. Given the crudeness of the approximation this is not a bad result. The reader is invited to show that for T near to, but below, T_c,

$$C_B = k\frac{2T - T_c}{T_c} \quad \text{and} \quad m = \gamma\left(\frac{T_c - T}{T_c}\right)^{1/2} \tag{14.12}$$

which are to be compared to the exact results (14.1) and (14.2). Again the agreement is fair, the magnetization going to zero as the $\frac{1}{2}$ power rather than the $\frac{1}{8}$ power. The specific heat does not possess a singularity at T_c. The problem with the mean field approximation is that it predicts spontaneous magnetization in all dimensions, whereas it is known that this does not exist in one dimension. The reason for this will become apparent later; it suffices to say that we have to account, at least approximately, for the fluctuations in the internal field.

14.4 The renormalization group and the one-dimensional Ising model

Underlying the renormalization group there is some very fundamental and profound physics. Unfortunately we do not have the space to go into this in detail and in any case it would take us far beyond the boundaries of an undergraduate text. What we shall present here is the straightforward calculational procedure and in doing so we will lean heavily on the presentation given by Maris and Kadanoff.[1] A major problem with the mean field approximation was that the result was unable to distinguish between the different effects which occur when one goes from one to two dimensions. In particular it predicted a phase transition in one dimension, which we know from the exact solution does not occur. Use of the renormalization group is an attempt to correct this result; we start therefore by briefly explaining the underlying philosophy of the method.

Suppose we start with an Ising model of N spins and that we can find a transformation that relates the (canonical) partition function of that system to the partition function for a system of $N/2$ spins. That is to say, if we could find the solution for the system with $N/2$ spins then we could have the solution for N spins. We could now repeat this procedure and find the partition function for $N/2$ spins in terms of that for $N/4$ spins and so on. Eventually we could reduce the problem to finding the partition function for just 2 spins (assuming all the time that N factorizes as an integral power of 2 – not a very restricting assumption). Now the 2-spin problem can easily be solved as in Section 7.4 and the procedure could be reversed, allowing the construction of the partition functions for 4 spins, 8 spins, and so on until we are back to the original problem of N spins. We would then have solved the problem. In practice, what we shall find is a set of iterative equations (between say n spins and $n/2$ spins) and then solve these to find the partition function for N spins.

To illustrate the method we consider a one-dimensional Ising model in the absence of an external field, the partition function for which can be written in the form

$$Q = \sum_{s_1 s_2 s_3 \ldots} \ldots \exp\left[K(s_1 s_2 + s_2 s_3)\right] \times \exp\left[K(s_3 s_4 + s_4 s_5)\right] \ldots \tag{14.13}$$

where

$$K = J/kT \tag{14.14}$$

is a dimensionless temperature-dependent coupling constant. Remembering that the only possible value of the s's are ± 1, we can perform the sum over all the even spins, with the result

$$Q = \sum_{s_1 s_2 s_3 \ldots} \ldots \{\exp[K(s_1 + s_3)] + \exp[-K(s_1 + s_3)]\}$$
$$\times \{\exp[K(s_3 + s_5)] + \exp[-K(s_3 + s_5)]\} \ldots \quad (14.15)$$

The problem now is that although we have eliminated $N/2$ of the spins, the expression for Q no longer has the form of the one-dimensional Ising model. However, if we could find a transformation such that, for example,

$$\exp[K(s_1 + s_3)] + \exp[-K(s_1 + s_3)] = f(K) \exp[K' s_1 s_3] \quad (14.16)$$

then (14.15) would be reduced to the partition function for the one-dimensional Ising model for $N/2$ spins with a coupling constant K', scaled by a factor $[f(K)]^{N/2}$. It is easy to find the form of the function $f(K)$ and the value of K'. If we first put $(s_5 = +1; s_2 = +1)$ in (14.16), and then $(s_5 = +1; s_2 = -1)$, we find the following two equations:

$$K' = \tfrac{1}{2} \ln \cosh(2K) \quad (14.17)$$

and

$$f(K) = 2 \cosh^{1/2}(2K) \quad (14.18)$$

Thus we can write

$$Q(N, K) = [f(K)]^{N/2} Q(N/2, K') \quad (14.19)$$

Remembering that the Helmholtz free energy $F = -kT \ln Q$ is extensive, i.e., proportional to N, it follows that $\ln Q$ is proportional to N so the quantity

$$\zeta = (1/N) \ln Q \quad (14.20)$$

is independent of N. Thus from (14.18) and (14.19),

$$\zeta(K') = 2\zeta(K) - \ln[2 \cosh^{1/2}(2K)] \quad (14.21)$$

This equation and (14.17) above form a set of iterative equations for the function $\zeta(K)$ so that, if we can solve them, we can proceed to find the values of $\zeta(K)$ for all values of K. Equation (14.20) then enables us to find the partition function for all values of the inverse temperature K and hence the thermodynamic properties of the system.

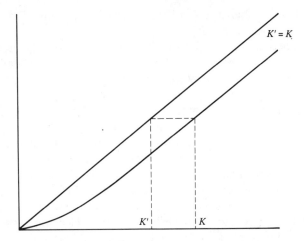

Figure 14.1 The right-hand side of (14.17) plotted as a function of K, together with the line $K' = K$. The dotted lines show how the iteration procedure always leads to a lower value of K'.

To find the solution, let us consider equation (14.17). Figure 14.1 shows the right-hand side of that equation plotted as a function of K and also the 45° line $K' = K$. The right-hand side for small values of K behaves like $K^2/8$ and for large K like K. Thus for a given K we generate a value of K' which is smaller than K, and as we continue the process we gradually converge upon $K = 0$. Fortunately we cannot go through that point, which would be physical nonsense (implying negative temperatures), since at $K = 0$ equation (14.17) has a solution $K' = K$. Such a point is known as a fixed point, and has considerable physical significance, as we shall see when we come to consider two dimensions. Thus this set of iterative equations should be used when starting from a high value of K. The inverses of (14.17) and (14.21) are easily obtained; they are

$$K = \tfrac{1}{2} \cosh^{-1}[\exp(2K')] \qquad (14.22)$$

and

$$\zeta(K) = \tfrac{1}{2} \ln 2 + \tfrac{1}{2}K' + \tfrac{1}{2}\zeta(K') \qquad (14.23)$$

Similar arguments to those given above show that (14.22) generates successively larger values of K, gradually converging on the fixed point at infinity. It turns out that these equations produce a more accurate result than the original iteration equation since, as K increases, any error in the initial value of $\zeta(K)$ is successively reduced with each iteration. For the original equations where K decreases with each iteration, any original error is steadily increased. Thus we shall use (14.22) and (14.23), once we have found a suitable starting value for $\zeta(K)$.

Remembering that K, defined by (14.14), is inversely proportional to the temperature, we need to find a value of ζ for a very small value of K, which corresponds to a very high temperature, i.e., $T \gg J/k$. At such temperatures the interaction between the spins is negligible and hence we have a completely random orientation of the spins. The corresponding value of $\zeta(K)$, as the reader can easily check by putting $K' = K$ in (7.61) and taking the limit $K' \to 0$, is ln 2. If we assume that kT is 100 times greater than J, then K has the value 0.01. We are now in a position to start the iteration. Question 14.1 asks the reader to carry this out. The agreement with the exact result is excellent. The iteration proceeds smoothly from one fixed point to the other. If there were a phase change then we would expect some discontinuity, such as would occur if there existed another fixed point at some finite but non-zero value of K. Then it would be impossible to proceed from say a small value of K to a large one, simply because the iteration would finish at the new fixed point.

Having demonstrated that the renormalization group method gives the correct result in one dimension we can now go on and look at the results in two dimensions. The situation is needless to say more complicated than in one dimension and we give here only the outline. Any details that the reader is unable to fill in for himself can be found in the paper of Maris and Kadanoff.[1]

14.5 The renormalization group and the two-dimensional Ising model

As in one dimension we write the canonical partition function in such a way that each alternate spin on a square lattice is shown coupled to its nearest neighbours. Figure 14.2 illustrates the point.

The partition function is then

$$Q = \sum_{s_1 s_2 s_3 \ldots} \ldots \exp[Ks_5(s_1 + s_2 + s_3 + s_4)]$$
$$\times \exp[Ks_6(s_2 + s_3 + s_7 + s_8)] \ldots \qquad (14.24)$$

If we now sum over the spins labelled 5, 6, etc., we get

$$Q = \sum_{s_1 s_2 s_3 \ldots} \ldots \{\exp[K(s_1 + s_2 + s_3 + s_4)]$$
$$+ \exp[-K(s_1 + s_2 + s_3 + s_4)]\} \ldots \qquad (14.25)$$

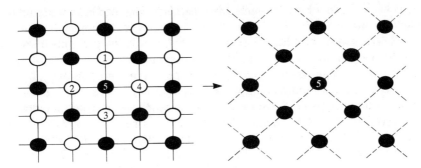

Figure 14.2 In two dimensions the summation is performed over the open circles. This leads to a new square lattice rotated through 45°. The nearest-neighbour separation is √2 times the original spacing, which presumably implies a weaker interaction. However, this is achieved at the expense of the more complicated partition function (14.26).

Thus, as in the one-dimensional case, we have managed to eliminate half the spins at the cost of altering the form of the partition function. However, the next step in the one-dimensional case does not have an exact analogue in two dimensions. The reader can easily verify that if he/she tries to write the analogue of (14.16). Because there are now four spins involved in each term, there are four conditions that have to be satisfied and only two variables K' and $f(K)$. It turns out that the best one can do is to write (14.25) in the form

$$Q = [f(K)]^{N/2} \sum_{s_1 s_2 \ldots} \exp\left(K_1 \sum_{pq} S_p S_q + K_2 \sum_{pq\,NNN} S_p S_q \right.$$

$$\left. + K_3 \sum_{pqrs\,SQR} S_p S_q S_r S_s \right) \tag{14.26}$$

where the first summation is taken over nearest neighbours, the second over next-nearest neighbours and the third over all spins forming a square surrounding any one spin. $f(K)$, K_1, K_2 and K_3 are given by

$$f(K) = 2\cosh^{1/2}(2K) \cosh^{1/8}(4K) \tag{14.27}$$

$$K_1 = \tfrac{1}{4} \ln \cosh(4K) \tag{14.28}$$

$$K_2 = \tfrac{1}{8} \ln \cosh(4K) \tag{14.29}$$

$$K_3 = \tfrac{1}{8} \ln \cosh(4K) - \tfrac{1}{2} \ln \cosh(2K) \tag{14.30}$$

Clearly we can no longer directly apply exactly the iterative procedure of the one-dimensional case. The new partition function on the right-hand side of (14.26) does not have the same form as the original partition function for the Ising model. The price that we have had to pay for removing half of the spins is a more complicated system with interactions not only between nearest neighbours but also next-nearest neighbours as well as groups of four spins. We must therefore resort to an approximation. The most obvious one is to ignore K_2 and K_3, simply on the physical grounds that these interactions should be weak compared to the nearest-neighbour interaction. However, if we do this, then it is easy to see that the equation which determines K_1 has the same form as that which determines K' in the one-dimensional case.[1]

Unless we are going to experience great difficulty we must put $K_3 = 0$, since this involves interactions between groups of four spins. The other two terms involve only interactions between pairs of spins. A possible approximation is to try and deal with the next-nearest neighbour terms by a mean field approximation, i.e., replace the effect of this term by an enhancement of the nearest-neighbour interaction. In the ground-state where all the spins are aligned, the total ground-state energy for the $N/2$ lattice is $-NkTK_1 + NkTK_2$. This suggests that we write

$$K' = K_1 + K_2 \qquad (14.31)$$

which, using (14.28) and (14.29), gives

$$K' = \tfrac{3}{8} \ln \cosh(4K) \qquad (14.32)$$

The first thing to note is that this equation has the form of the equation discussed in the footnote† and satisfies the condition for an additional finite non-zero fixed point. In fact putting $K' = K$ yields a fixed point at $K_c = 0.506\,98$, which is to be contrasted with the correct result of $0.440\,69$. The other recurrence relation is found in a similar way to the one-dimensional case; the result is

$$\zeta(K') = 2\zeta(K) - \ln[2\cosh^{1/2}(2K)\cosh^{1/8}(4K)] \qquad (14.33)$$

By drawing a diagram similar to Figure 14.1 it is easy to see that the iteration always effects a shift away from the fixed point, which is therefore known as an unstable fixed point.

Question 14.1 guides the reader through the calculation of the heat capacity in the neighbourhood of this fixed point. The result is

---

† The reader should become convinced that the replacement in (14.17) of the 2 by 4 to give (14.28) makes no difference to the overall shape of the curve in Figure 14.1. Thus there are only fixed points at zero and infinity. In fact it is not difficult to see that equations of the form $K' = (1/a) \ln \cosh(bK)$ only have an additional non-zero finite fixed point if $b > a$.

$$C \propto |T - T_c|^{0.131} \tag{14.34}$$

This is to be contrasted with the correct result given by Onsager which predicts a logarithmic singularity at the phase change.

We now conclude our discussion of the use of the renormalization group method in the theory of second-order phase transitions. Interested readers should refer to the paper by Maris and Kadanoff[1] and the references cited therein.

14.6 Monte Carlo methods

In the above sections we have referred to the evaluation of the partition function for the Ising model by using numerical methods. In this section we give one of the more well-known methods for carrying out such a calculation. The object of this section is not to obtain the precise numbers, but to encourage readers to perform the calculations for themselves. In particular we give at the end of this chapter a simple Basic program to evaluate and display some of the properties of a finite two-dimensional Ising model. By experimenting with this program and possibly modifying it the reader should gain some insight, not only into the properties of the Ising model but also into nature of phase transitions.

We denote a particular state of the Ising model by the set of numbers $\{s_j\}$, that is to say, we specify the state of each spin, whether it is up, $s_j = 1$, or down, $s_j = -1$. The corresponding energy will be denoted by $E(\{s_j\})$ and the average of any quantity $M(\{s_j\})$ using the canonical distribution by $\langle M \rangle$. The principle behind the Monte Carlo method is that one selects a series of different spin configurations, such that the number of times a particular configuration appears is consistent with the canonical distribution. Averages of particular quantities are then found by simply calculating their value for every configuration generated, summing all these values and then dividing by the total number of configurations. The problem we have to solve is therefore: how do we generate successive configurations such that the number of times a particular configuration is generated is consistent with the canonical distribution?

Let us start with an arbitrarily chosen configuration $\{s_j\}$. We can generate a new configuration by choosing one of the spins at random and reversing it. We can continue this process for as long as we like, each reversal giving a new configuration. At the end we can count up how many times a particular configuration say $\{s'_j\}$ occurs and then divide by the total number of configurations we have generated; this should then give us a measure of the probability that the particular state $\{s'_j\}$ will occur. The probabilities calculated in this way will all be equal and will not be given by the canonical distribution. The reason for this is not hard to see; in making the transition from one configuration to another we have assumed that the probability that this will occur is independent of the particular transition involved. In other words, the transition probabilities (see Section 2.2) are all equal. However, we know from the conditions of detailed balance, which ensures that the resulting distribution is consistent with the canonical distribution (the reader should consult any textbook on quantum mechanics to find the derivation of this result), that the ratio of transition probabilities must satisfy

$$W(\{s_j\}:\{s_j'\})/W(\{s_j'\}:\{s_j\}) = \exp\{-[E(\{s_j'\} - E(\{s_j\})]/kT\} \tag{14.35}$$

where $W(\{s_j\}:\{s_j'\})$ is the probability that the system makes a transition from $\{s_j\}$ to $\{s_j'\}$. One possibility for fulfilling this condition is to choose

$$W(\{s_j'\}:\{s_j\}) \propto 1 \tag{14.36}$$
$$\text{if } [E(\{s_j'\}) - E(\{s_j\})] < 0$$

and

$$W(\{s_j\}:\{s_j'\}) \propto \exp\{-[E(\{s_j'\}) - E(\{s_j\})]/kT\} \tag{14.37}$$
$$\text{if } [E(\{s_j'\}) - E(\{s_j\})] > 0$$

If we were to choose such transition probabilities then we know, because of the condition for detailed balance, that we would obtain a canonical distribution of states. However, we still do not know, when changing a given configuration by reversing an arbitrary spin, whether we should accept the given change or not. If the energy change is negative then the above algorithm tells us that we must always accept the transition. When the energy change is positive, according to (14.37), sometimes we should accept it and sometimes not; in a large number of trials of the same transition, the ratio of acceptances to rejections should be in the ratio of the right-hand side of (14.37). That of course does not help us in deciding whether to accept or reject a particular transition. However, it turns out, and the statistically-minded reader is invited to prove it, that if we generate a random number x, in the range $0 < x < 1$, for each transition and then accept the transition if the right-hand side of (14.37) is greater than x and reject it if the converse is true, we will generate a transition probability consistent with (14.37).

A computer program to perform a Monte Carlo calculation of this kind would be structured as follows. Starting from a given configuration, the computer selects at random an arbitrary spin and reverses its direction. It then calculates the resulting energy change and if that is negative, then the new configuration is accepted. If the energy change is positive, the program generates a random number in the range 0 to 1 and compares the right-hand side of (14.38) with that number. If it meets that acceptance criterion, then again the new configuration is accepted and the physical quantity to be averaged is calculated for the configuration. If the acceptance criterion is not met then the reversed spin is put back to its original value and no contribution is made to the computed quantities. The Basic program given at the end of this chapter performs this algorithm for a two-dimensional square lattice with sides $N\%$, $M\%$. The output consists of two parts: a graphical display showing the spin configuration changing with time and a calculation of the magnetization. For the initial configuration the user has two choices: either the case where the state of each spin is chosen at random or where there is a block of up spins and a block of down spins. The reader can learn much physics, as well as becoming familiar with the Monte Carlo method, by running this program a number of times under different conditions.

In particular, if one selects the initial configuration of blocks of spins (referred to in the program as 'phase separated') and a low temperature, say 0.5, with zero magnetic field, one will observe that the 'phases' remain separated, apart from fluctuations at the boundary. Now it is an exact result that as $T \to 0$, the state of the system is one in which *all* the spins are lined up. In terms of the blocks of spins this would require one whole block to be reversed. The Monte Carlo method, as given above, only reverses one spin at a time and although in principle the state of nearly all the spins being aligned could be reached by reversing one spin at a time, the probability of this happening is vanishingly small. To get to the known ground state one would have to introduce the possibility of going from one configuration to the other by reversing a large *block* of spins. This example illustrates how the Monte Carlo method can lead to the 'trapping' of the system in a particular region of configuration space. This can lead to difficulties if the initial choice of configuration is one that has a low probability of occurrence (given by the canonical distribution) at the chosen temperature.† One method of getting over this problem is known as window sampling and to illustrate the technique we take a particular example.

Suppose that for our given lattice we wanted to calculate the probability that the system had a particular magnetization at a given temperature. Clearly in order to do this we would have to collect our statistics from the whole range of possible configurations, which would mean that we would have to prevent the system being trapped in a particular region of

† (The subject of phase transitions is fraught with difficulties, which are of course part of its fascination. The statement that at $T = 0$ the equilibrium state is the ground state with all the spins aligned is strictly true, but it is less certain whether the system can approach that state in practice. Ferromagnets are known to possess domains, that is, regions whose magnetizations are internally uniform but whose directions of magnetization vary discontinuously from domain to domain. Although such a state is not the ground state, its structure does determine the temperature dependence of the entropy as $T \to 0$ consistently with the third law.)

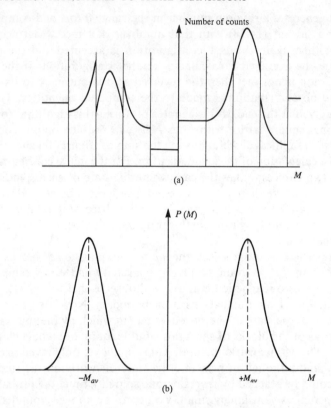

Figure 14.3 (a) A schematic count of a typical 'window sampling' count of the number of times a particular magnetization is observed in a Monte Carlo calculation. (b) The smooth curve derived from (a) by rendering it continuous and normalizing its area to unity.

configuration space. Thus, in the case discussed above where we start from a phase-separated initial state, the probability of having a magnetization around zero (if we start from roughly equal numbers of up and down spins) would be very large and the probability that we would have a magnetization around its maximum value would be extremely small. This is in complete contrast to the correct result where we expect the opposite to be true. The reason as we stated above is that because of the limitations of the Monte Carlo method it is very easy to get 'trapped' in a particular region of configuration space.

We have therefore to ensure that the system visits all possible regions of configuration space. We can do this for the case of the probability of the system having a particular value of the magnetization, by dividing the magnetization up into a number of windows. Thus, if the possible values of the magnetization lie between $-M$ and $+M$, then we can choose a window say ΔM, such that $R\Delta M = 2M$, where R is an integer. If $M = 400$ then typically $\Delta M = 10$ and thus we would have $R = 80$ windows. The trick then is to start from a configuration which is inside a particular window, for example, the magnetization lying in the range 250 to 260; the program is then as above, except there is an additional criterion for any newly generated configuration, namely, if the magnetization lies outside of the boundaries of the window, then that configuration is rejected. The modified program counts the number of times each allowed configuration is visited. After a sufficient number of trials, to ensure that one has reasonable statistics, the program moves to the adjacent window and repeats the process. In this way the entire spectrum of possible configurations is visited.

One can easily calculate the probability that a particular magnetization will occur within a particular window. The results are shown schematically in Figure 14.3(a). The reason that the curves are not continuous is simply that we have not allowed the system to move smoothly from one set of configurations to the other; however, if we assume, not unreasonably, that the probability function must be continuous, then we can simply adjust the height of the curve within each window to ensure that it joins onto the curve in the next window. We can fix the absolute values by ensuring that the total area under the curve is unity. To get a smooth curve in such a calculation requires a very large number of trials within each window; this means that it is not really practical to write a program in Basic and one really needs either a program in say Fortran or assembler language. For that reason we do not give a typical program. For more information the reader is invited to consult one of the number of excellent texts available on computer calculations in physics.

Figure 14.3(b) shows a schematic representation of such a calculation. The probability is doubly peaked, unlike the form of the probabilities that we have discussed up until now. Both peaks are narrow and we have to conclude that the system is equally likely to be in $+M_{av}$ or $-M_{av}$. Which of these two macroscopic states it is in, is entirely determined by the previous history of the sample, that is, the direction of a vanishingly small magnetic field. Such a curve illustrates the phenomenon of broken symmetry. In the absence of a magnetic field, considerations of symmetry tell us that the expected magnetization should be zero and consequently the probability should in fact peak at that value. In fact, the symmetry is broken as we have illustrated. As the temperature is increased the two peaks move closer together until, at the critical temperature, they combine at $M = 0$. However, at that temperature, the width of the peak is large giving rise to fluctuations in the magnetization, consistent with our earlier discussion. This phenomenon of broken symmetry is by no means confined to the simple Ising model but we shall discuss it no further here.

14.7 Phase changes and dimensionality

We conclude this chapter with a short discussion of the effects on the quantum-mechanical states of the Ising model in different dimensions and the effects on the presence or absence of a phase transition. Many of the effects discussed here can be illustrated with the computer model given at the end of the chapter, and again the reader is urged to experiment with it.

In the non-interacting case we could specify the energy by a single parameter $N_\uparrow$ with a corresponding degeneracy $g_{N\uparrow}$. When the interactions are present we will see that this is no longer the case, in that not only are the energy levels shifted but also the degenerate energy levels are split. To see this, consider the ground state. Even in the presence of interactions this is still the state when all the spins are parallel to the field i.e., all the $s_i = 1$. Thus

$$E_0 = -N\gamma B - \tfrac{1}{2}NzJ \tag{14.38}$$

It is still singly degenerate and is lower than the non-interacting ground state by the term $\tfrac{1}{2}NzJ$. As in the mean field approximation, it will prove useful to interpret this energy as that of a set of N spins parallel to both an external field B and an internal field B_{int} given by

$$B_{int} = zJ/\gamma \tag{14.39}$$

The ground-state energy can now be written

$$E_0 = -N\gamma B - \tfrac{1}{2}N\gamma B_{int} \tag{14.40}$$

To obtain the first excited state energy we reverse one of the spins. The increase in energy due to the reversed spin in the external field is $2\gamma B$; the increase due to the coupling is found as follows. Reversing a spin means that we lose z bonds, each of which has an energy $-J$ and gain z bonds each with energy $+J$. Thus the net gain in energy from the interactions is $2zJ$. The first excited state energy is then

$$E_1 = -(N-2)\gamma B - \tfrac{1}{2}(N-4)zJ \tag{14.41}$$

As for the non-interacting case, this state is N-fold degenerate, since we can reverse any of the N spins. It is also possible to interpret this energy in terms of an external and an internal field. For the $(N - z - 1)$ spins not including the reversed spin or its z neighbours, the field they see is the external field plus the internal field given by (14.39). So their contribution to the energy is

$$-(N-z-1)\gamma B - \tfrac{1}{2}(N-z-1)zJ \tag{14.42}$$

The reversed spin also sees the same field, so its contribution to the energy is

$$\gamma B + \tfrac{1}{2} z J \tag{14.43}$$

The change of sign is of course due to the fact that the spin is reversed. However, those spins which are nearest neighbours to the reversed spins see a different field because one of their neighbours is reversed. This localized internal field is

$$B_{int} = (z - 2)J/\gamma \tag{14.44}$$

Thus their contribution to the energy is

$$-z\gamma B - \tfrac{1}{2} z(z - 2)J \tag{14.45}$$

Adding these three contributions together gives (14.41). The point to notice here is that the internal field seen by the neighbours of the reversed spin is less than that of the other spins. Thus it becomes energetically favourable to reverse a second spin which is a neighbour of an existing reversed spin, rather than to reverse any other spin. Thus, the second excited state will be formed by reversing a nearest neighbour spin to the already reversed spin. It follows that the $\tfrac{1}{2}N(N-1)$ states of the non-interacting case, with energies $-(N-4)\gamma B$, will be split into $\tfrac{1}{2}N(N-1-z)$ states with energies

$$-(N - 4)\gamma B - \tfrac{1}{2} z J(N - 8) \tag{14.46}$$

which corresponds to reversing two uncoupled spins and $Nz/2$ states with energies

$$-(N - 4)\gamma B - \tfrac{1}{2} z J(N - 8) - 2zJ \tag{14.47}$$

The last term is due to the fact that the two reversed spins' mutual interaction contributes $-J$ to the energy rather than J. We conclude, therefore, that it will be energetically favourable for the system to form isolated pockets of reversed spins, rather than in the non-interacting case where this ordered state is not to be preferred over the more random state, where there exist only isolated reversed spins. For low-lying energies there will be far more of these latter states but for the higher energies this will no longer be true. We should also note for later reference that the average internal field is decreasing as we increase the energy.

From the point of view of statistical mechanics we are led to the conclusion that the effect of the interactions is to produce a more ordered state than in the non-interacting case. Thus, we would expect the entropy to be lower and possibly the presence of spontaneous magnetization (note that the internal field does not vanish when the external field is zero). It also shows that we would expect localized fluctuations in the magnetization and our previous discussions suggest that when the fluctuations become macroscopic we should expect a phase change. We have observed above that the average internal field decreases with increasing energy and thus we would expect it to fall with increasing temperature and eventually be zero. Since at zero temperature it has the value zJ, we might expect this to happen with $T \sim zJ/k$.

Finally in this section we give a qualitative explanation of the difference between one and two dimensions.[2] Consider a cluster of βN spins all anti-parallel to the field, the remaining $(1 - \beta)N$ spins surrounding the cluster being parallel to the field (where $0 \leqslant \beta \leqslant \frac{1}{2}$). Now there is an interface energy between the cluster and its surroundings. In three dimensions this is equal to $(\beta N)^{2/3} J$, in two dimensions $(\beta N)^{1/2} J$, and in one dimension $2J$. Thus, in two and three dimensions, states in which the total magnetization is close to zero (or equal to zero if $\beta = \frac{1}{2}$) have high energies compared with the ground state. In one dimension, however, the corresponding state is only $2J$ above the ground state, which means that states corresponding to magnetizations close to zero are readily accessible. Further, there are $\frac{1}{2}N$ such states (if $\beta = \frac{1}{2}$) so it is not just one pathological isolated state that gives zero magnetization. Thus, in one dimension we would expect that the maximum magnetization of the ground state would disappear as the temperature is raised above J/Nk and the external field is allowed to go to zero. The possibility for spontaneous magnetization in one dimension therefore does not exist and consequently a phase change does not occur.

CHAPTER 14: SUMMARY OF MAIN POINTS

1. The Ising model of interacting spins is a reasonably tractable model, which can be adapted to apply to physical situations other than magnetism.
2. It has exact solutions in one and two dimensions, although the method of solution is mathematically difficult, particularly in two dimensions.
3. The Ising model does not exhibit a phase change in one dimension but does in two. The exact solution in these two dimensions provides a check for any approximate solutions.
4. The simplest approximation is the mean field approximation, which shows a phase change in all three dimensions.
5. Application of the renormalization group to one dimension enables the exact solution to be found by iteration.
6. In two and three dimensions, the renormalization group cannot be used to obtain the exact solution; an approximate solution does, however,

yield a phase change in two dimensions. The behaviour of the specific heat in the region of the phase change is given by a power law, rather than the logarithmic singularity predicted by the exact result.
7. An alternative to the renormalization group is to use Monte Carlo methods, which depend upon the high speeds of modern computers.
8. Starting from a given configuration of spins, the computer generates a large number of allowable configurations. The macrosopic properties are calculated for each configuration. The thermodynamic properties are then averaged over all configurations generated.
9. The Monte Carlo method generates a double-peaked distribution as the temperature is reduced to zero, corresponding to the possibilities of all the spins being 'up' or all the spins 'down'.
10. Finally, the question of the relationship between dimensionality and the existence or not of a phase change, is related to the number of neighbours in the different dimensions.

QUESTIONS

14.1 Perform the iteration of equations (14.22) and (14.23), starting with $K = 0.01$ and $\zeta(K) = \ln 2$. Compare your results with the exact result calculated from

$$Q = [2\cosh(K)]^N$$

or

$$\zeta(K) = \ln[2\cosh(K)]$$

14.2 Show that the approximation (14.32) has a fixed point as $K \to 0$ ($T \to 0$); but as $K \to \infty$ ($T \to 0$), the iteration does not converge to a fixed point. What are the physical reasons for these results?

14.3 If equations (14.32) and (14.33) are iterated they lead to increasing errors in $\zeta(K)$. Show that the inverses of these equations are

$$K = \tfrac{1}{4}\cosh^{-1}[\exp(\tfrac{8}{3}K')]$$

and

$$\zeta(K) = \tfrac{1}{2}\zeta(K') + \tfrac{1}{2}\ln 2 + \tfrac{1}{6}K' + \tfrac{1}{4}\ln\{\cosh^{-1}[\exp(\tfrac{8}{3}K')]\}$$

Iterate these equations and show that in this case the error decreases and $\zeta(K)$ converges on the value 1.215 653 as K approaches K_c from either side.

14.4 Assume that the function $\zeta(K)$ has a non-analytic part, which behaves like

$$2a|K - K_c|^{2-\alpha}$$

Next expand K', regarded as a function of K, around K_c and then compare terms on either side of (14.33). Hence show that

$$\alpha = 2 - \frac{\ln 2}{\ln\left(\frac{dK'}{dK}\bigg|_{K=K_c}\right)}$$

Evaluate this expression by using (14.32) and use standard statistical mechanics to obtain (14.34).

14.5 Run the Basic program given at the end of the chapter for various temperatures at zero external field, until you are satisfied that the mean value of the magnetization is steady. Hence plot the magnetization as a function of temperature. Identify the critical temperature and then repeat the process for temperatures close to that temperature so that you can identify the nature of the singularity. Repeat the calculation to find the effect of size on the result.

14.6 Choose the option where there is no interaction between the spins and then compare the magnetization with the theoretical value. How does the root mean square deviation compare with the theoretical value? Again, what is the effect of size on the results?

The following is a simple Basic program to evaluate and display some of the properties of a finite two-dimensional Ising model.

```
HIDEMOUSE

GRAFRECT 0,0,SCREENWIDTH,SCREENHEIGHT
CLG 0
TXTEFFECTS %00000001
TXTRECT 0,0,SCREENWIDTH,SCREENHEIGHT
PRINT TAB(20,10)"Two dimensional Ising model"
PRINT TAB(20,12)"Monte Carlo Calculation"
PRINT TAB(20,14)"Inverse critical temperature is ~0.5"
PRINT TAB(20,16)"The appropriate range of values of the
    magnetic field"
PRINT TAB(20,18)"is from zero to say 4"
PRINT TAB(20,26)"Press any key to continue."
D$=GET$
TXTEFFECTS %00000000
```

```
GRAFRECT 0,0,SCREENWIDTH,SCREENHEIGHT
CLG 0
CLS

\ Dimensions of rectangle
INPUT TAB(0,15)"Dimensions of the rectangle M%<60
    N%<25." M%,N%
DIM A(M%+2,N%+2)

INPUT TAB(0,20) "Temperature" T
K=1/T
INPUT"Value of the magnetic field" h
IF h< >0 THEN PROCcouple ELSE J=1
IF h< >0 AND J=0 THEN PROCexact

\ Initialise
icount=0:jcount=1:mag=0

\ Choose initial configuration
REPEAT
PRINT "Initial configurations are random up and down spins
    or phase separated
PRINT"Choose initial configuration. r/p"
S$=GET$
UNTIL S$="p" OR S$="r" OR S$="R" OR S$="P"
IF S$="p" OR S$="P" THEN PROCphase
IF S$="r" OR S$="R" THEN PROCrandom
PRINT TAB(20,8)"        magn."
PRINT TAB(10,8)"      configs."
PRINT TAB(10,40)"Accepted configs","        ","average
    mag","rms"
IF h< >0 AND J=0 THEN PRINT TAB(42,10)"Exact
    magnetisation = ";magexact%

\ Changes configuration
REPEAT
PROCflip
PROCtest
IF accepted THEN
PROCboun
PROCupdate
jcount=jcount+1
summag=summag+mag
avmag=summag/jcount
IF avmag=0 THEN BEEP
```

```
        summagsq = summagsq + mag^2
        IF avmag< >0 THEN rms = SQR(summagsq/jcount −
          SQUARE(avmag))/ABS(avmag) ELSE rms = 0
      ELSE
        A(m,n) = − A(m,n): \ new configuration not accepted
        avmag = 0
      ENDIF
      icount = icount + 1
      IF icount MOD 250 = 0 THEN
        PRINT TAB(10,10)icount
        PRINT TAB(20,10)    mag
        PRINT TAB(10,42)    jcount,"           ",FORMAT$(avmag,
          "ZZZ.D") "          ",FORM (rms,"ZD.DDDD")
      ENDIF
    UNTIL FALSE
    SHOWMOUSE
END


DEFPROCcouple
PRINT "Do you want the spins coupled or not? y/n"
B$ = GET$
IF B$ = "Y" OR B$ = "y" THEN J = 1 ELSE J = 0
ENDPROC


DEFPROCexact
y = EXP(K*h)
tnh = (y − 1/y)/(y + 1/y)
magexact% = INT(N%*M%*tnh)
ENDPROC


DEFPROCrandom
\ Initial display
CLS
LOCAL sum
sum = 0
FOR I% = 1 TO M%
  FOR J% = 1 TO N%
    IF RND(1) < 0.5 THEN A(I%,J%) = − 1 ELSE A(I%,J%) = 1
    sum = sum + A(I%,J%)
  NEXT J%
NEXT I%
mag = sum
IF mag< >0 THEN
  summag = mag
```

```
summagsq = SQUARE(mag)
ELSE
summag = 1
summagsq = 1
ENDIF
PROCbocon
PROCdisplay
ENDPROC

DEFPROCphase
\ Phase separated
INPUT "Number of rows before interface >0 & <M%" R%
CLS
LOCAL sum
sum = 0
FOR I% = 1 TO M%
FOR J% = 1 TO N%
IF I%<=R% THEN A(I%,J%)=1 ELSE A(I%,J%)=-1
sum = sum + A(I%,J%)
NEXT J%
NEXT I%
IF mag<>0 THEN
summag = mag
summagsq = SQUARE(mag)
ELSE
summag = 1
summagsq = 1
ENDIF
PROCbocon
PROCdisplay
ENDPROC

DEFPROCbocon
FOR I% = 1 TO M%
A(I%,0) = A(I%,N% + 1)
A(I%,N% + 1) = A(I%,1)
NEXT I%
FOR I% = 1 TO N%
A(0,I%) = A(M%,I%)
A(M% + 1,I%) = A(1,I%)
NEXT I%
ENDPROC

DEFPROCdisplay
\ Initial display
```

```
CLS
FOR I% = 1 TO M%
FOR J% = 1 TO N%
IF A(I%,J%) = 1 THEN FILLSTYLE 1,1 ELSE FILLSTYLE
  0,0
CIRCLE I%*8 + 100,J%*8 + 100,4
NEXT
NEXT
ENDPROC

DEFPROCflip
IF M%⟩1 THEN m = RND(M%) ELSE m = 1
IF N%⟩1 THEN n = RND(N%) ELSE n = 1
A(m,n) = -A(m,n)
ENDPROC

DEFPROCtest
deltae = -A(m,n)*J*(A(m-1,n) + A(m+1,n) + A(m,n-1)
  + A(m,n+1)) - 2*h*A(m,n)
X = RND(1)
accepted = (EXP(-deltae*K) > X)
ENDPROC

DEFPROCboun
IF n = 1 THEN A(M% + 1,n) = -A(M% + 1,n)
IF n = N% THEN A(0,n) = -A(0,n)
IF m = 1 THEN A(m,N% + 1) = -A(m,N% + 1)
IF m = M% THEN A(m,0) = -A(m,0)
ENDPROC

DEFPROCupdate
IF A(m,n) = 1 THEN FILLSTYLE 1,1 ELSE FILLSTYLE 0,0
CIRCLE m*8 + 100,n*8 + 100,4
mag = mag + 2*A(m,n)
ENDPROC
```

REFERENCES

1   Maris H.J. and Kadanoff L.P., *Am. J. Phys.* (1978), **46**, 652.
2   An exact, but mathematically very sophisticated, solution for the two-dimensional Ising model with finite field has been conjectured by A.B. Zamolodchikov of the Landau Institute for Theoretical Physics in Moscow. A preprint RAL-89-001, entitled *Integrals of Motion and S-matrix of the (scaled) $T = T_c$ Ising Model with Magnetic Field* (January 1989), is available from the Rutherford Appleton Laboratory, Chilton, Didcot, Oxon., OX11 0QX, UK.

Appendices

Appendix 1
The method of Lagrange or undetermined multipliers

It happens in statistical mechanics (and indeed often elsewhere – for instance in commerce and engineering design, and even in urban planning) that one needs to identify the maximum or minimum value of a function of several variables under the condition that there exist one or more fixed relationships – constraints – between the variables. An example of this is to find the dimensions of the rectangular box of greatest volume which can be made out of a fixed area of cardboard. Mathematically it would be necessary to maximize the function $V(a,b,c) = abc$ subject to the constraint $A(a,b,c) = 2(ab+bc+ca)$. This is left as an exercise in Question A1.1. In more general terms the problem might be stated as follows.

Maximize

$$f(x_1, x_2, \ldots x_i, \ldots x_N) \tag{A1.1}$$

subject to the constraints

$$g_j(x_1, x_2, \ldots x_i, \ldots x_N) = 0 \tag{A1.2}$$

where $j = 1, 2, \ldots M$. Clearly the problem is overdetermined unless $M < N$. The extremum (i.e., maximum or minimum) of the function f is characterized by

$$df = \sum_{i=1}^{N} \left(\frac{\partial f}{\partial x_i}\right) dx_i = 0 \tag{A1.3}$$

If there were no constraints, all the dx_i could be independently varied at the extremum but, because of the constraints, g_j, there is a free choice of only $(N - M)$ of them. Expressing (A1.2) in the differentiated form

$$dg_i = \sum_{i=1}^{N} \left(\frac{\partial g_i}{\partial x_i}\right) dx_i = 0 \qquad (A1.4)$$

and combining with (A1.3) yields

$$df - \sum_{j=1}^{M} \alpha_j dg_j = 0 \qquad (A1.5)$$

where the α_j can be *any* numerical constants, and are called undetermined or Lagrange multipliers. Then, from equations (A1.3-5),

$$\sum_{i=1}^{N} \left[\left(\frac{\partial f}{\partial x_i}\right) - \sum_{j=1}^{M} \alpha_j \left(\frac{\partial g_j}{\partial x_i}\right) \right] dx_i = 0 \qquad (A1.6)$$

This is the sum of N terms and we are completely free to choose all the M values of α_j and also $(N - M)$ of the dx_i. In fact we make a particular choice of the α_j in such a way that the first M terms in square brackets are to be zero (although we do not need at this stage to know just what the numerical choices are - those values emerge later). For the remaining $(N - M)$ terms in the summation (A1.6) the dx_i can be independently varied because the constraints have been taken care of. It follows that each remaining term in square brackets must be independently zero in order that the sum for any choices of the dx_i be zero. There are thus $(N-M)$ equations of the form

$$\left(\frac{\partial f}{\partial x_i}\right) - \sum_{j=1}^{M} \alpha_j \left(\frac{\partial g_j}{\partial x_i}\right) = 0 \quad \text{for} \quad i = (M+1), \ldots, N \quad (A1.7)$$

The net result of this manoeuvre is that all N terms in square brackets of (A1.6) may be set independently to zero, the first M of them because we have deliberately made suitable notional choices of the α_j, and the remaining $(N - M)$ because of the arguments leading to equation (A1.7). Thus, there is in principle no barrier to the solution (A1.6) for all the x_i which maximize f, in terms of the α_j, all of which now follow the constraint equations (A1.2). The problem is now solved in general. Usually, however, we only need to deal with cases in which M is only one or two (for example, fixed mean energy and mean number of particles in the grand canonical distribution) so that only one or two Lagrange multipliers appear.

We now illustrate the method by using the procedure to maximize $f(x, y) = xy$ subject to the constraint $x^2 + y^2 = 1$, using equation numbers which parallel those in the general approach presented above. (This particular problem would of course be readily solved more conventionally by substituting for $y^2 = 1 - x^2$ and maximizing $f(x) = x(1 - x^2)^{\frac{1}{2}}$. Naturally the result is the same.) We maximize

$$f(x, y) = xy \tag{A1.8}$$

subject to the constraint

$$g(x, y) = x^2 + y^2 - 1 = 0 \tag{A1.9}$$

Hence, in differential form,

$$df = xdy + ydx = 0 \tag{A1.10}$$

at the maximum, and

$$dg = 2xdx + 2ydy = 0 \tag{A1.11}$$

Combining these equations gives

$$df - \alpha dg = 0 \tag{A1.12}$$

where α is the single Lagrange multiplier. Hence

$$(y - 2\alpha x)dx + (x - 2\alpha y)dy = 0 \tag{A1.13}$$

We now choose α so that the first bracket is zero. That is, $\alpha = y/2x$, where of course the values of x and y are those at the maximum. Since the first bracket has been set at zero, it follows from (A1.13) that the second must also be zero.

$$x = 2\alpha y = \frac{y^2}{x} \quad \text{so} \quad x = \pm y$$

Thus, from (A1.2a), $x = 2^{-\frac{1}{2}}$. Also,

$$\alpha = \frac{x}{2y} = \tfrac{1}{2}$$

QUESTIONS

A1.1 Find the dimensions of the rectangular box of greatest volume which can be made out of a fixed area A of cardboard.

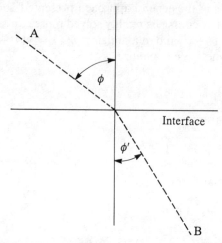

A1.2 In an application of Lagrange's method of undetermined multipliers, consider points A and B in two different media as shown in the figure below. Show that the time required for light to propagate from A to B is minimized if

$$\frac{v}{v'} = \frac{\sin \varphi}{\sin \varphi'}$$

where v is the speed of light above the interface side and v' is the speed of light below the interface. In optics this minimization of propagation time is known as Fermat's Principle.

Appendix 2
Number fluctuations in the grand canonical distribution

Direct differentiation of the grand canonical partition function, equation (4.55), shows that

$$\langle N^2 \rangle - \langle N \rangle^2 = \frac{\left\{ \sum_{N=0}^{\infty} N^2 \sum_n \sum_j \exp\left[\frac{(\mu N - E_n)}{kT} \right] \right\}}{\Xi}$$

$$- \frac{\left\{ \sum_{N=0}^{\infty} N \sum_n \sum_j \exp\left[\frac{(\mu N - E_n)}{kT} \right] \right\}^2}{\Xi^2}$$

$$= (kT)^2 \left[\frac{\partial^2 \ln \Xi}{\partial \mu^2} \right]_{T,V} = kT \left(\frac{\partial \mathcal{N}}{\partial \mu} \right)_{T,V} \quad \text{(A2.1)}$$

where we have used the result (4.70),

$$\mathcal{N} = kT \left(\frac{\partial \ln \Xi}{\partial \mu} \right)_{T,V}$$

In order to manipulate equation (A2.1) into a more recognizable form, we can also use the identity

$$\left(\frac{\partial \mathcal{N}}{\partial \mu} \right)_{T,V} = \frac{\left(\frac{\partial p}{\partial \mu} \right)_{T,V}}{\left(\frac{\partial p}{\partial \mathcal{N}} \right)_{T,V}} \qquad (A2.2)$$

Differentiating the Gibbs free energy

$$G = \mu \mathcal{N} = U + pV - TS$$

and using the result

$$\mu d\mathcal{N} + TdS = dU + pdV$$

we find

$$VdP = SdT + \mathcal{N}d\mu \qquad (A2.3)$$

Therefore the numerator of equation (A2.2) is

$$\left(\frac{\partial p}{\partial \mu} \right)_{T,V} = \frac{\mathcal{N}}{V} \qquad (A2.4)$$

and the denominator is found by starting from the bulk modulus as follows.

$$B = -V \left(\frac{\partial p}{\partial V} \right)_{T,\mathcal{N}} = -\mathcal{N} \left(\frac{\partial \mu}{\partial V} \right)_{T,\mathcal{N}} \qquad (A2.5)$$

The Helmholtz free energy is

$$F = U - TS$$

and so we find

$$dF = -pdV + \mu d\mathcal{N} - SdT$$

which leads to the Maxwell relation

$$\left(\frac{\partial p}{\partial \mathcal{N}}\right)_{T,V} = -\left(\frac{\partial \mu}{\partial V}\right)_{\mathcal{N},T} \tag{A2.6}$$

Combining (A2.5) and (A2.6) gives

$$B = \mathcal{N}\left(\frac{\partial p}{\partial \mathcal{N}}\right)_{T,V} \tag{A2.7}$$

and hence the denominator of equation (A2.2) is identified. This result plus equations (A2.1), (A2.2) and (A2.4) leads directly to equation (4.79).

Appendix 3
Proof that, for independent distinguishable sub-systems, $Q = \prod_i q_i$

Consider an assembly of distinguishable and independent sub-systems, each with a different label i. Let each sub-system have its own private range of states with energies $\epsilon_{i,s(i)}$. Any microstate of the assembly may be represented diagrammatically, and the figure below shows an example of such a representation for a six-subsystem assembly.

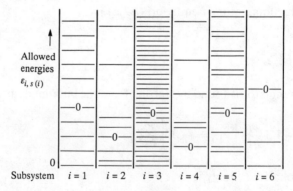

The total energy associated with the above microstate is the sum

$$\epsilon_{1,4} + \epsilon_{2,3} + \epsilon_{3,11} + \epsilon_{4,1} + \epsilon_{5,5} + \epsilon_{6,2} \tag{A3.1}$$

More generally we shall write $E_{\{set\}}$ where $\{set\}$ denotes a set of pairs of numbers i (specifying the sub-system) and $s(i)$ (specifying the state which the sub-system is in). For the microstate illustrated above, the set is:

i	$s(i)$
1	4
2	3
3	11
4	1
5	5
6	2

Now

$$Q = \sum_{\text{sets}} \exp(-E_{\{\text{set}\}}/kT) \quad \text{(by definition)} \tag{A3.2}$$

$$= \sum_{\text{sets}} \exp\left(-\sum_i \epsilon_{i,s(i)}/kT\right) \quad \text{(independence)} \tag{A3.3}$$

$$= \sum_{\text{sets}} \prod_i \exp(-\epsilon_{i,s(i)}/kT) \tag{A3.4}$$

where the last step is simply the replacement of the exponential of a sum by a product of exponentials. The product

$$\prod_i \exp(-\epsilon_{i,s(i)}/kT)$$

has one factor for each sub-system. For the microstate illustrated and tabulated above, the product is

$$\exp(-\epsilon_{1,4}/kT) \times \exp(-\epsilon_{2,3}/kT) \times \exp(-\epsilon_{3,11}/kT)$$
$$\times \exp(-\epsilon_{4,1}/kT) \times \exp(-\epsilon_{5,5}/kT) \times \exp(-\epsilon_{6,2}/kT)$$

The meaning of the sum over all sets is that such a term must be added for each distinct microstate, and each has the form of a string of multiplied exponentials (as for the example above). Every possible combination must be included. But clearly every possible combination can alternatively be produced by first summing over all possible $s(i)$ values for a given sub-system, and then multiplying all the sums together, i.e.,

$$Q = \sum_{\text{sets}} \prod_i \exp(-\epsilon_{i,s(i)}/kT) \tag{A3.5}$$

$$= \prod_i \sum_{s(i)} \exp(-\epsilon_{i,s(i)}/kT) \tag{A3.6}$$

But

$$q_i = \sum_{s(i)} \exp(-\epsilon_{i,s(i)}/kT) \tag{A3.7}$$

is none other than the sub-system partition function. Therefore

$$Q = \prod_i q_i \quad \text{as promised} \tag{A3.8}$$

Appendix 4
Tables of the chemical potential and energies for Fermi and Bose gases

For the purposes of comparison of Bose and Fermi gases it is convenient to introduce a new characteristic temperature T_0, given by

$$T_0 = \frac{h^2}{2mk} \left(\frac{N}{2\pi Vg}\right)^{2/3} \tag{A4.1}$$

In terms of T_0, the Fermi degeneracy temperature $T_f = \mu_0/k$ and the Bose degeneracy temperature T_b are, using equations (10.33) and (10.59),

$$T_f = \left(\frac{3}{2}\right)^{2/3} T_0 = 1.310\,371\, T_0 \tag{A4.2}$$

$$T_b = \frac{1}{\pi} \left(\frac{2\pi}{2.612}\right)^{2/3} T_0 = 0.571\,408\, T_0 \tag{A4.3}$$

The equation for μ for fermions and for bosons for $T \geqslant T_b$ (remembering that $\mu = 0$ in integrals for $T < T_b$) can be written in the form

$$1 = \tau^{3/2} \int_0^\infty x^{\frac{1}{2}} \left\{ \exp\left(x - \frac{\mu^*}{\tau}\right) \pm 1 \right\}^{-1} dx \tag{A4.4}$$

where $\tau = T/T_0$ and $\mu^* = \mu/kT_0$.

Similarly the internal energies can be expressed in terms of the parameters τ and μ^*. For fermions

$$u_f^*(\tau) = \frac{U_f}{\mathcal{N}kT_0} = \tau^{5/2} \int_0^\infty x^{3/2} \left\{ \exp\left(x - \frac{\mu_f^*}{\tau}\right) + 1 \right\}^{-1} dx \quad \text{(A4.5)}$$

For bosons we have, for $T > T_b$ or $\tau > 0.571\,408$,

$$u_b^*(\tau) = \frac{U_b}{\mathcal{N}kT_0} = \tau^{5/2} \int_0^\infty x^{5/2} \left\{ \exp\left(x - \frac{\mu_b^*}{\tau}\right) - 1 \right\}^{-1} dx \quad \text{(A4.6)}$$

and for $T < T_b$ or $\tau < 0.571\,408$,

$$u_b^*(\tau) = \frac{U_b}{\mathcal{N}kT_0} = \tau^{5/2} \int_0^\infty x^{5/2} [\exp(x) - 1]^{-1} dx = 1.783\,292\,\tau^{5/2}$$

(A4.7)

Computed values of μ_f^*, μ_b^*, u_f^* and u_b^* are given in the table below.

τ	μ_f^*	μ_b^*	u_f^*	u_b^*
0.00	1.310 37	0.000 00	0.786 22	0.000 00
0.10	1.304 04	0.000 00	0.804 88	0.005 64
0.20	1.284 19	0.000 00	0.858 49	0.031 90
0.30	1.248 25	0.000 00	0.939 45	0.087 91
0.40	1.194 11	0.000 00	1.038 86	0.180 46
0.50	1.121 43	0.000 00	1.150 39	0.315 25
0.60	1.030 92	−0.001 70	1.270 09	0.494 67
0.70	0.923 72	−0.031 42	1.395 54	0.678 63
0.80	0.801 02	−0.091 97	1.525 17	0.855 08
0.90	0.663 98	−0.177 69	1.657 94	1.026 47
1.00	0.513 63	−0.284 68	1.793 13	1.194 24
1.10	0.350 91	−0.410 15	1.930 24	1.359 33
1.20	0.176 67	−0.551 92	2.068 88	1.522 35
1.30	−0.008 35	−0.708 31	2.208 78	1.683 75
1.40	−0.203 48	−0.877 95	2.349 73	1.843 83
1.50	−0.408 14	−1.059 71	2.491 56	2.002 84
1.60	−0.621 79	−1.252 65	2.634 13	2.160 95
1.70	−0.843 95	−1.455 96	2.777 34	2.318 30
1.80	−1.074 18	−1.668 94	2.921 10	2.475 01
1.90	−1.312 10	−1.890 99	3.065 34	2.631 16
2.00	−1.557 34	−2.121 57	3.210 01	2.786 83
3.00	−4.351 78	−4.812 44	4.672 05	4.326 55
4.00	−7.735 56	−8.034 51	6.149 21	5.850 00
5.00	−11.288 71	−11.645 54	7.633 56	7.365 94
6.00	−15.238 45	−15.564 19	9.121 97	8.877 67
7.00	−19.435 92	−19.737 50	10.612 96	10.386 78
8.00	−23.846 09	−24.128 19	12.105 69	11.894 11
9.00	−28.442 59	−28.708 55	13.599 66	13.400 18
10.00	−33.204 86	−33.457 18	15.094 55	14.905 32

τ	μ_f^*	μ_b^*	u_f^*	u_b^*
20.00	−87.367 13	−87.545 55	30.066 89	29.933 08
30.00	−149.357 57	−149.503 26	45.054 62	44.945 36
40.00	−216.438 40	−216.564 56	60.047 30	59.952 69
50.00	−287.306 20	−287.419 04	75.042 31	74.957 68
60.00	−361.192 58	−361.295 58	90.038 62	89.961 37
70.00	−437.589 57	−437.684 93	105.035 76	104.964 23
80.00	−516.136 02	−516.225 22	120.033 45	119.966 54
90.00	−596.561 86	−596.645 96	135.031 53	134.968 46
100.00	−678.657 41	−678.737 20	150.029 92	149.970 07

QUESTION

A4.1 Explain why the tables above are not especially useful for metals. Write and run a computer program to tabulate μ_f^* and u_f^* as functions of τ for $0 < \tau < 0.1$, using equations (A4.4) to (A4.7). With the chemical potentials found, the internal energies can be obtained by using a standard routine for numerical integration (some pocket calculators can do this, if rather slowly). The method for first finding the chemical potentials is necessarily more elaborate however. For each value of τ, one has to make a guess (or informed choice) of μ and then evaluate the integral in equation (A4.4) numerically. The result, multiplied by $\tau^{2/3}$, must be compared with unity and the estimate of μ adjusted and the integration repeated until finally, usually after many iterations, a value of μ emerges which satisfies equation (A4.4) within an acceptable margin. This is how the tables above were generated.

Appendix 5
Integrals for the Fermi gas

We are concerned with evaluating integrals of the form

$$I = \int_0^\infty K(\epsilon) \left\{ \exp\left[\frac{\epsilon - \mu}{kT}\right] + 1 \right\}^{-1} d\epsilon \qquad \text{(A5.1)}$$

in the limit as $T \to 0$. $K(\epsilon)$ is an arbitrary function of ϵ which is smooth near $\epsilon = \mu$. If we put $K(\epsilon) = dG(\epsilon)/d\epsilon$ and integrate by parts, then

$$I = -G(0) - \int_0^\infty G(\epsilon) \frac{dn}{d\epsilon} d\epsilon \qquad \text{(A5.2)}$$

where

$$n(\epsilon) = \left\{ \exp\left[\frac{\epsilon - \mu}{kT}\right] + 1 \right\}^{-1} \quad \text{(A5.3)}$$

Now functions $n(\epsilon)$ and $dn/d\epsilon$ have the general appearances shown in the figures below.

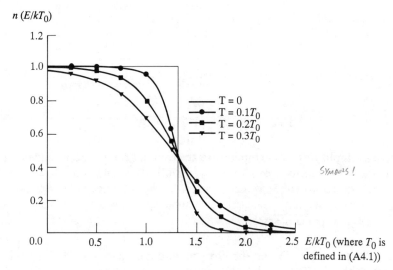

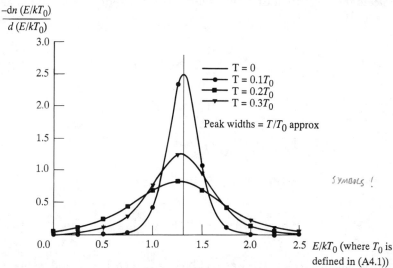

Thus $dn/d\epsilon$ is greatest for values of ϵ close to μ and vanishes rapidly if $|\epsilon - \mu|$ is greater than kT. We therefore expand $G(\epsilon)$ in a Taylor series about the point $\epsilon = \mu$ and obtain for the integral

$$I = -G(0) + G(\mu) + G'(\mu)n_1 + \frac{1}{2!}G''(\mu)n_2$$

$$+ \frac{1}{3!}G'''(\mu)n_3 + \ldots \qquad (A5.4)$$

where the primes denote derivatives with respect to ϵ, and $n_1, n_2, \ldots$ the moments given by

$$n_r = -\int_0^\infty (\epsilon - \mu)^r \frac{dn}{d\epsilon} d\epsilon \qquad (A5.5)$$

The substitution $(\epsilon - \mu)/kT = x$ leads to

$$n_r = +(kT)^r \int_{-\infty}^\infty x^r \frac{1}{(1+e^x)(1+e^{-x})} dx \qquad (A5.6)$$

where a simplifying low-temperature approximation has been made by setting $-\mu/kT$ equal to $-\infty$ in the lower limit of the integral. Now

$$\frac{1}{(1+e^x)(1+e^{-x})}$$

is an even function of x and therefore all odd moments vanish. For the even moments the coefficient of $(kT)^r$ is just a number. For the second moment we have

$$n_2 = (kT)^2 \int_{-\infty}^\infty \frac{x^2}{(1+e^x)(1+e^{-x})} dx = \frac{\pi^2}{3}(kT)^2 \qquad (A5.7)$$

Thus

$$I = -G(0) + G(\mu) + \frac{\pi^2}{6}G''(\mu)(kT)^2 + 0(T^4) \qquad (A5.8)$$

or, in terms of the original function K

$$I = \int_0^\mu K(\epsilon) d\epsilon + \frac{\pi^2}{6}K'(\mu)(kT)^2 + 0(T^4) \qquad (A5.9)$$

In particular, if $K(\epsilon) = (2\pi Vg/h^3)(2m)^{3/2}\epsilon^{\frac{1}{2}}$, then $I = \mathcal{N}$ and we have

$$\mathcal{N} = \frac{4\pi Vg}{3h^3}(2m\mu)^{3/2} + \frac{Vg\pi^3}{6h^3}(2m)^{3/2}\mu^{-\frac{1}{2}}(kT)^2 \qquad (A5.10)$$

This equation has to be inverted to find μ as a function of $\mathcal{N}/V$. As a first approximation we drop the term in T^2 and find

$$\mu_0 = \frac{h^2}{8m}\left(\frac{6\mathcal{N}}{\pi Vg}\right)^{2/3} \tag{A5.11}$$

which agrees with equation (10.32). To find the second approximation we substitute μ_0 given by (A5.11) into the second term on the right-hand side of (A5.10) and after rearrangement find

$$\mu^{3/2} = (2m)^{-3/2}\frac{6\mathcal{N}h^3}{8\pi Vg} - (kT)^2\frac{\pi^2}{8}(2m)^{3/2}\left(\frac{8\pi Vg}{6\mathcal{N}h^3}\right)^{\frac{1}{2}} \tag{A5.12}$$

$$= \mu_0^{3/2}\left\{1 - (kT)^2\frac{\pi^2}{8}\mu_0^{-2}\right\} \tag{A5.13}$$

Since the second term in this expression is proportional to T^2, we can use the binomial expansion to find the 2/3 root and obtain, finally,

$$\mu = \mu_0 - \frac{\pi^2}{12}\mu_0^{-1}(kT)^2 \tag{A5.14}$$

Returning to equation (A5.9), if $K(\epsilon) = (2\pi Vg/h^3)(2m\epsilon)^{3/2}$ then $I = U$, so that

$$U = \frac{4\pi Vg}{5h^3}(2m)^{3/2}\mu^{5/2} + \frac{3\pi^3 Vg}{6h^3}(2m)^{3/2}\mu^{\frac{1}{2}}(kT)^2 \tag{A5.15}$$

Now in this expression, in the first term on the right-hand side we substitute for μ using (A5.14) and in the second term we use (A5.11) for μ, since this is already of order T^2. After rearrangement we find

$$\frac{U}{N} = \frac{3}{5}\mu_0 + \frac{\pi^2}{4}\frac{(kT)^2}{\mu_0} \tag{A5.16}$$

Numerical answers to questions

Chapter 3

3.2 (a) 1.099 (b) 1.011 (c) 1.386 (d) 0.736
3.4 (i/a) 0.500 (i/b) 0.693 (ii/a) 0.167 (ii/b) 1.792 (iii/a) 0.148 (iii/b) 3.090
3.5 $P_0 = 0.5$, $P_1 = P_{-1} = 0.25$
3.6 (b) $N_0 = 752$, $N_1 = 197$, $N_2 = 51$

Chapter 4

4.7 $\langle N_0 \rangle = 4.018$, $\langle N_\epsilon \rangle = 0.816$, $\langle N_{2\epsilon} \rangle = 0.166$

Chapter 5

5.4 $c/k = 1.069$
5.7 $T/\theta_t = 3.185 \times 10^5$

Chapter 6

6.4 $T \ll (2h/k)(\sigma/3m)^{1/2} = 13\,\text{K}$
6.7 Debye with $\theta_D = 346.0\,\text{K}$ gives specific heats of $16.50\,\text{J kg}^{-1}\,\text{deg}^{-1}$ at 15 K and $3.886\,\text{J kg}^{-1}\,\text{deg}^{-1}$ at 35 K; Einstein with $\theta_E = 261.3\,\text{K}$ gives specific heats of $16.50\,\text{J kg}^{-1}\,\text{deg}^{-1}$ at 115 K and $0.797\,\text{J kg}^{-1}\,\text{deg}^{-1}$ at 35 K.
6.8 2158 K

Chapter 7

7.7 2.87 K; 4.16 mK
7.8 $8.87 \times 10^{-8}\,\text{m}^3$

Chapter 9

9.5 $s_v = 187.3 \text{ J mol}^{-1}$; $s_l = 89.6 \text{ J mol}^{-1}$

9.8 $\theta_v = 3275 \text{ K}$; $\nu = 6.82 \times 10^{13} \text{ Hz}$

9.9 $\theta_r = 86 \text{ K}$ gives the ortho to para ratio as 0.029 at 30 K, and 3 at $T \gg 86 \text{ K}$

9.11 $\theta_v = 3.10 \text{ K}$ is much less than 300 K so $C_V = \tfrac{5}{2} Nk$ to a good approximation.

9.13 $T^{3/2} \exp[-(1.57 \times 10^5)/T] = 10^4 \alpha^2/[9.65(1-\alpha)]$ which, for $\alpha = 0.01$, gives $T = 9790 \text{ K}$. Yes, because the degeneracy temperatures for the electrons, protons, and atoms are respectively 0.65 K, 0.35 mK, and 7.65 mK, all of which are much lower than 9790 K.

9.14   (b) $\alpha = 6 \times 10^{-8}$

Chapter 10

10.4   (d) 79   (e) 3.6 MeV   (f) about 10 000 MeV

Chapter 11

11.2   (a) 10 eV   (b) 10^8 eV   (c) 10^{-3} eV

11.3 $1.27 \times 10^9 \text{ Pa}$

11.4   (a) 2.3×10^{-3}%   (b) 0.358 nm

11.5 $\mu_0 = 3.25 \text{ eV}$. Between 0 K and 300 K the fall in μ is indicated in equation (A5.13) by the term $(\pi kT)^2/8\mu_0^2 = 7.6 \times 10^{-5}$ so the change is certainly very small.

11.6   The Debye lattice specific heat is calculated with $\theta_D = 385 \text{ K}$ to be $22.9 \text{ J mol}^{-1} \text{deg}^{-1}$ at 300 K. The electronic contribution is calculated from equation (10.39) with $T_f = 1.347 \times 10^5 \text{ K}$ to be $9.138 \times 10^{-2} \text{ J mol}^{-1} \text{deg}^{-1}$ at 300 K. The ratio is 0.4 %.

11.7   (a) 0.02; 1.4×10^{-17} (b) $1.3 \times 10^3 \text{ K}$ (c) $4.4kT$

11.8   (a) $4.5 \times 10^{20} \text{ W}$ (b) $2.2 \times 10^{15} \text{ rad sec}^{-1}$ (c) $3.2 \times 10^{-9} \text{ Pa}$

11.11   385 K; $9.90 \times 10^5 \text{ Pa}$

11.12 241 m sec^{-1}

11.13   (a) $8.23 \times 10^{-23} \text{ J}$ (b) 5.9 K   (c) $-6.3 \times 10^{-22} \text{ J}$ (d) $7.11 \times 10^5 \text{ Pa}$; $2 \times 10^6 \text{ Pa}$

Chapter 12

12.2 $kT/v_A^0 = 2.54 \times 10^7 \text{ Pa}$; $kT/v_B^0 = 2.64 \times 10^7 \text{ Pa}$. These magnitudes ensure that the exponential factors in equation (12.54) are close to unity for pressures of the order of the given pure vapour pressures ($9.4 \times 10^5 \text{ Pa}$ and $6.9 \times 10^4 \text{ Pa}$). $3.3 \times 10^5 \text{ Pa}$.

12.8 $4.49 \times 10^3 \text{ Pa}$ corresponding to a liquid helium-4 column of height 3.16 m.

Index

acessibility of quantum states 2
adiabatic perturbation 12–13
adsorption 216–219, 242
  localized 217–219
  mobile 217–219
  sites 217
adsorption isotherms 217–219
  of argon on $MgBr_2$ 225
  of helium-4 on porous glass 220
  see also two-dimensional
alloys 204
amplitude of oscillation see modes of vibration
anti-symmetric wave function 118–119
argon 1–2, 77–78, 94–98, 235, 239

BET see Brunauer-Emmett-Teller
black-body radiation see photons
Bohr magneton 104
Bose–Einstein condensation 157, 160
Bose gas 144–148, 157–163
boson 119
boundary conditions 2
Boyle temperature 239
bridge equation
  canonical 43
  grand canonical 48
  microcanonical 38
Brunauer-Emmett-Teller (BET) equation for multimolecular adsorption 219
bulk modulus 39, 52, 188, 274

canonical distribution 40–47
characteristic temperature for
  Bose degeneracy 158, 278
  classical gas criterion 122–129
  Debye solid 89
  Einstein solid 76
  Fermi degeneracy 151, 278
  paramagnetic crystal 105
  rotation 65
  translation 67, 123–124
  vibration 64
chemical potential 17, 192–197
  of Bose gas 152, 159, 224, 278–280
  in canonical distribution 193
  equality in phases in equilibrium 197
  of Fermi and Bose gases compared 152
  of Fermi and Bose two-dimensional gases compared 224
  of Fermi gas 151–152, 224, 278–280, 283
  and Gibbs free energy 193
  and relation to latent heat 199
  for two-dimensional Fermi and Bose gases 167
classical statistical mechanics 230–234
cohesive energy 1, 173
coin tossing 26
configuration integral 236
contact potential 214–216
correspondence principle 231
coupling constant 249

286 INDEX

Curie law
  electron gas 213
  classical assembly of magnetic moments 111, 228

Dalton's law of partial pressures 132, 210
Debye functions 89, 94
Debye solid 87–90
Debye temperature 89
degeneracy 42, 48
  magnetic 104
  see also characteristic temperature
density of momentum states 176
density of energy states
  classical gas 148
  Debye solid 88–89
  diamond 86
  particle-in-a-box 124
  photons 179
  oscillating one-dimensional model 79–84
  oscillating three-dimensional model 85–87
  two-dimensional gaseous layer 223
  see also frequency spectrum
dice throwing 27–29
diffusion in a solid 8
dimensionality and phase changes 260–262
dissociation 132–135
  degree of 134
distinguishable particles 116–118
distributions
  canonical 40–47
  choice of 60
  constant pressure 55
  grand canonical 47–54
  Maxwell-Boltzmann 124
  microcanonical 35–40
distribution function for radiation 181
Dulong and Petit law 77, 93

effective mass 171
eigenvalues 11, 14
Einstein model of solids 29–31, 40, 58, 75–79
  vapour pressure 197–199

Einstein temperature 76
elastic constant see Hooke's constant of elasticity
electric polarization 111
electric susceptibility 111
electron gas in metals 170–174
  binding energy 173
  heat capacity 171–172
  jellium model 172–173
  magnetic susceptibility 211–213
  paramagnetism 211–213
  pressure 172
electronic canonical partition function of a molecule 61–63, 70
electromagnetic radiation
  canonical partition function 179
  grand canonical partition function 180
  internal energy 180
  photon approach 179–180
  pressure 180
  wave approach 178–179
energy
  fractional deviation in canonical distribution 44–45
  internal see internal energy
  mean 16
enthalpy 196
  minimization in relation to equilibrium between phases 196
entropy
  of Bose gas 147, 162
  of Bose systems 147
  in canonical distribution 43
  of Einstein solid 76
  of Fermi gas 147, 155
  of Fermi system 147
  in grand canonical distribution 50, 109
  of interacting gas 244
  in microcanonical distribution 38
  of mixing 132, 137–138
  of monatomic gas 129
  in one-dimensional oscillating lattice model 83
  of paramagnetic crystal in canonical distribution 105
  of paramagnetic crystal in grand canonical distribution 109

in three-dimensional oscillating
lattice model 86
and uncertainty 38, 43, 49
equation of state 14
equilibrium constant 135
equilibrium in two phases 197
ethane 142
exclusion principle *see* Pauli exclusion
principle

Fermi energy 187
see also chemical potential for Fermi
gases
Fermi gas 144–148, 150–155
relativistic 166–167
Fermi integrals 281–283
fermion 119, 145
fluctuations
critical 53
in Fermi system 155–157
see also fractional deviations
fractional deviations in
Bose system 163–164
energy in canonical distribution
44–45
energy of an Einstein solid 79
Fermi system 155–157
magnetization 112
number in grand canonical
distribution 52
frequency spectrum 85
of Debye solid 88–89
of three-dimensional oscillating
lattice model 85
of two-dimensional Debye solid 221
see also density of energy states

gas (classical) 122–143, 148–150
of electric dipoles 139, 242
hydrogen, deuterium, and tritium 28
monatomic, thermodynamic
properties 128–129
polyatomic, thermodynamic
properties 129–131
of rods, one-dimensional 244
also see Maxwell–Boltzmann gas
gas (interacting) 230–241
thermodynamic properties 238–241
gas mixtures

non-reacting 132
reacting 132–135
Gibbs free energy 194
of Bose gas 161
minimization of 196
Gibbs' paradox 132, 137–138
grand canonical distribution 47–54,
144–148
gyromagnetic ratio 104

Hamiltonian 2, 11, 113–114, 118
heat bath, system in contact with
40–47
heat and particle bath, system in
contact with 47–54
heat capacity 15
of Bose gas 160, 176
of classical anharmonic oscillator
243
of Debye solid 89–90
in 1-D and 2-D Debye models 93,
221
of Einstein solid 76–78
of electron gas 171
of Fermi gas 154
of Fermi and Bose gases compared
154
of helium-3 adsorbed on porous
glass 222
of liquid helium-3 184–185
of liquid helium-4 175–177
of monatomic gas 129
of NO and HD 130–131
of non-reacting gas mixtures 132
of oscillating one-dimensional lattice
model 84
of oscillating three-dimensional
lattice model 86
of paramagnetic solid in the
canonical distribution 106
of paramagnetic solid in the grand
canonical distribution 109
of photons 180
of polyatomic gas 130
of solid argon 78
of solid potassium 172
of spin waves in ferromagnetic solid
94
of surface waves 92

helium-3, liquid 183
helium-3/helium-4 liquid mixtures 183–186
  degeneracy temperature 184
  equilibrium concentrations 206–209
  heat capacity 184–185
  magnetic susceptibility 213
  osmotic pressure 184–185
  phase separation 203–208
helium-4, liquid 174–178
  excitation spectrum 177
  heat capacity 175–177
  internal energy 176–177
  lambda line 206
  phase diagram 174, 206
  phonons and rotons 176
Helmholtz free energy 193
  minimization 196
Hooke's constant of elasticity 79–80
hydrogen, ortho- and para- 140

ideal mixture 199–203
  in microcanonical distribution 200
imperfect gas *see* gas (interacting)
indistinguishable particles 118–120
information theory 19–31
interaction energy 1, 133, 235
internal energy 3
  of Bose gas 147, 160, 278–280
  of Bose system 147
  of Debye solid 89–90
  in canonical distribution 44
  of Einstein solid 77
  of electron gas 171, 173
  of Fermi gas 147, 152–153, 278–280, 283
  in grand canonical distribution 51
  of liquid helium-3/helium-4 mixtures 184
  of liquid helium-4 177
  of monatomic gas 131
  in microcanonical distribution 39
  minimization of 195
  of paramagnetic solid in the canonical distribution 106
  of photons 180
internal magnetic field 247, 260

Ising model 246–247
  see also renormalization group
isothermal bulk modulus *see* bulk modulus

jellium model *see* electron gas in metals

Lagrange multipliers *see* undetermined multipliers
lattice vacancy concentration 57–58

macroscopic state and description 10–11
magnetic susceptibility 111, 213
magnetic systems
  in canonical distribution 101–102
  in grand canonical distribution 102–103
  themodynamic functions 101–103
magnetization
  of electron gas 168, 213
  of paramagnetic solid in canonical distribution 105–106
  of paramagnetic solid in grand canonical distribution 109
many-particle states 113–115
Maxwell–Boltzmann distribution 124, 136, 148–150
Maxwell–Boltzmann gas 136, 148–150
  of electric dipoles 139
  in microcanonical distribution 39–40
mean field approximation 246–249
microcanonical distribution 35–40
microscopic state and description 2
mixtures *see* ideal mixtures
  see also helium-3/helium-4 liquid mixtures; alloys
modes of vibration 80, 82
  longitudinal and transverse 88
monatomic gas 128–129
Monte Carlo methods 255–259

normal mode coordinates 82
nuclear model 166
number of particles
  fractional deviation in grand canonical distribution 52, 272–274

INDEX 289

mean
  in Fermi and Bose systems 146
  in Bose gas 157, 159
  in Fermi gas 151
  in grand canonical distribution 47
  variable 16–17

occupation numbers
  mean
    in Bose and Fermi systems 146–147
    in Fermi gas 153
oscillating lattice model
  in one dimension 79–85
  in three dimensions 85–87
osmotic pressure 208–211
  of helium-3/helium-4 liquid mixtures 184–186

paramagnetism 100–112
partial pressures
  of liquid mixtures 201–203
  of non-reacting gas mixtures 132
particles
  distinguishable 116–118, 120
  indistinguishable 118–120
partition function
  for Bose system 145–146
  classical 233
  for classical gas 125–126
  canonical 42
  for diatomic molecule 70
  for Einstein solid 76
  for electronic levels 61
  factorization of 70–71, 75, 116–118, 123
  for Fermi system 145–148, 152–153
  grand canonical 48
  for independent distinguishable subsystems 275–277
  for magnetic subsystem 102
  microcanonical 36
  for monatomic gas 125–126
  for non-reacting gas mixture 132
  for oscillating one-dimensional lattice model 83
  for oscillating three-dimensional lattice model 85
  for paramagnetic crystal in canonical distribution 105
  for paramagnetic crystal in grand canonical distribution 108
  for particle in a box 67–69
  for polyatomic gases 129
  for rotator 64–67
  for simple harmonic oscillator 63–67
  for two molecules 70–71
Pauli exclusion principle 115, 119, 153
perturbation theory, time-dependent 12, 13
phase diagram
  of liquid helium-3/helium-4 mixtures 205–209
  of liquid helium-4 174
  of copper/silver alloy 204
phase integral 233
phase of oscillation see modes of vibration
phase separation 203–208
  equilibrium concentration 205–208
  of helium-3/helium-4 mixtures 183, 205–208
phase transitions 245–262
phonons 91
  in Debye solid 90–91, 189–190
  in liquid helium-4 176
photons 9, 180–182, 189
  thermodynamic properties 180
Planck's law of radiation 180–182
plasma 141
polyatomic gas 129–131
potassium (solid) heat capacity 172
pressure 14
  of Bose gas 147–148, 160
  of Bose system 147
  in canonical distribution 43
  of electron gas 172
  of Fermi system 147
  of Fermi gas 147–148, 152–153, 184
  in grand canonical distribution 51
  in microcanonical distribution 40
  of monatomic gas 129
  of non-reacting gas mixtures 132
  in oscillating three-dimensional model 87
  of photons 180
  of polyatomic gas 129–130
  see also partial pressures

probability 5
  of microstate 6
  of a system having a particular
    pressure 4
  see also distributions

quantum number 11
  magnetic 111

radial distribution function 243
radiation 178-183
  thermodynamic properties 180
  see also photons
  see also electromagnetic radiation
Raoult's law 203
renormalization group 249-255
  and the one-dimensional Ising model
    249-252
  and the two-dimensional Ising model
    252-255
reversible changes 12-13
root mean square deviation 5-6
  of energy in canonical distribution
    44-45
  of particle number in grand
    canonical distribution 52
rotational canonical partition function
  of a molecule 64-67, 70
rotator, rigid linear 64-67
rotons 176
rubber elasticity 8, 57

scale of uncertainty 21-22
  for equal probabilities 22
  rational requirements for 22
  for unequal probabilities 24-25
Schrödinger equation 2
second virial coefficient 238-240,
  242
  of argon gas 239
simple harmonic oscillator (SHO)
  63-64
solid
  Debye 87-90
  Einstein 29-31, 40, 58, 75-79
  one-dimensional oscillating lattice
    model 79-85
  paramagnetic 8, 100-109
  three-dimensional oscillating lattice
    model 85-87

specific heat see heat capacity
spin degeneracy 124
statistical inference 2, 19-34
Stefan-Boltzmann law 180-182
strategy of least bias 25-26
substrate 216
sub-systems, non-interacting 59-73,
  113-121
superleak 196, 211
surfaces 216-217
symmetric wave functions 118-119

thermodynamics
  first law 15, 17
  second law 37
  third law 62, 65
translational canonical partition
  function of a molecule 67-70
two-dimensional
  solid layers 219-223
  gaseous layers 223-225

uncertainty 2, 20-25
  function 23-25
  law of 31
  scale of 21-22
  maximization of 25
undetermined multipliers 269-271

vacancies in a crystal lattice 57-58
Van der Waals' equation 56, 240
Vant' Hoff's law 184, 186, 210
vapour pressure of
  Debye solid 228
  Einstein solid 197-199
  ideal liquid mixture 203
  liquid argon/krypton mixture 227
vibrational canonical partition function
  of a molecule 63-64, 70
virial coefficients 238
  second 238-241, 242
virial expansion for imperfect gas
  238

wave packet 126
Wien's law 180-182
work function 214, 227

zero-point energy 91, 179